State-of-the-Art Program on Compound Semiconductors 58 (SOTAPOCS 58)

Editors:

J. H. He

C. O'Dwyer

F. Ren

E. Douglas

C. Jagadish

S. Jang

Y.-L. Wang

R. Lynch

T. Anderson

J. Hite

Sponsoring Division:

 Electronics and Photonics

Published by
The Electrochemical Society

65 South Main Street, Building D
Pennington, NJ 08534-2839, USA

tel   609 737 1902
fax   609 737 2743
www.electrochem.org

ecstransactions ™

Vol. 69, No. 14

This book has been registered with Copyright Clearance Center.
For further information, please contact the Copyright Clearance Center,
Salem, Massachusetts.

Published by:

The Electrochemical Society
65 South Main Street
Pennington, New Jersey 08534-2839, USA

Telephone 609.737.1902
Fax 609.737.2743
e-mail: ecs@electrochem.org
Web: www.electrochem.org

ISSN 1938-6737 (online)
ISSN 1938-5862 (print)
ISSN 2151-2051 (cd-rom)

ISBN 978-1-62332-321-9 (CD-ROM)
ISBN 978-1-60768-679-8 (PDF)

Printed in the United States of America.

Preface

The papers included in this issue of *ECS Transactions* were originally presented in the symposium "State-of-the-Art Program on Compound Semiconductors 58 (SOTAPOCS 58)", held during the 228th meeting of The Electrochemical Society, in Phoenix, AZ, from October 11 to October 15, 2015.

***ECS Transactions*, Volume 69, Issue 14**
State-of-the-Art Program on Compound Semiconductors 58 (SOTAPOCS 58)

Table of Contents

Chapter 4
Nitride Materials & Devices

Chapter 5
Poster Session

Chapter 6
Radiation Effects

Chapter 7
Group IV Materials and Devices

Facts about ECS

The Electrochemical Society (ECS) is an international, nonprofit, scientific, educational organization founded for the advancement of the theory and practice of electrochemistry, electronics, and allied subjects. The Society was founded in Philadelphia in 1902 and incorporated in 1930. There are currently over 7,000 scientists and engineers from more than 70 countries who hold individual membership; the Society is also supported by more than 100 corporations through Corporate Memberships.

The technical activities of the Society are carried on by Divisions. Sections of the Society have been organized in a number of cities and regions. Major international meetings of the Society are held in the spring and fall of each year. At these meetings, the Divisions and Groups hold general sessions and sponsor symposia on specialized subjects.

The Society has an active publication program that includes the following:

Journal of The Electrochemical Society — (JES) is the leader in the field of electrochemical science and technology. This peer-reviewed journal publishes an average of 550 pages of 85 articles each month. Articles are published online as soon as possible after undergoing the peer-review process. The online version is considered the final version and is fully citable with articles assigned specific page numbers within specific issues. The date of online publication is the official publication date of record.

Journal of Solid State Science and Technology — (JSS) is one of the newest peer-reviewed journals from ECS launched in 2012. JSS covers fundamental and applied areas of solid state science and technology including experimental and theoretical aspects of the chemistry and physics of materials and devices. Articles are published online as soon as possible after undergoing the peer-review process. The online version is considered the final version and is fully citable with articles assigned specific page numbers within specific issues. The date of online publication is the official publication date of record.

Electrochemistry Letters — (EEL) is one of the newest journals from ECS launched in 2012. It is dedicated to the rapid dissemination of peer-reviewed and concise research reports in fundamental and applied areas of electrochemical science and technology. Articles are published online as soon as possible after undergoing the peer-review process. The online version is considered the final version and is fully citable with articles assigned specific page numbers within specific issues. The date of online publication is the official publication date of record.

Solid State Letters — (SSL) is one of the newest journals from ECS launched in 2012. It is dedicated to the rapid dissemination of peer-reviewed and concise research reports in fundamental and applied areas of solid state science and technology. Articles are published online as soon as possible after undergoing the peer-review process. The online version is considered the final version and is fully citable with articles assigned specific page numbers within specific issues. The date of online publication is the official publication date of record.

Electrochemical and Solid-State Letters — (ESL) was the first rapid-publication electronic journal dedicated to covering the leading edge of research and development in the field of solid-state and electrochemical science and technology. ESL was a joint publication of ECS and IEEE Electron Devices Society. Volume 1 began July 1998 and contained six issues, thereafter new volumes began with the January issue and contained 12 issues. The final issue of ESL was Volume 16, Number 6, 2012. Preserved as an archive, ESL has since been replaced by SSL and EEL.

Interface— *Interface* is an authoritative yet accessible publication for those in the field of solid-state and electrochemical science and technology. Published quarterly, this four-color magazine contains technical articles about the latest developments in the field, and presents news and information about and for members of ECS.

ECS Meeting Abstracts— *ECS Meeting Abstracts* contain extended abstracts of the technical papers presented at the ECS biannual meetings and ECS-sponsored meetings. This publication offers a first look into the current research in the field. ECS Meeting Abstracts are freely available to all visitors to the ECS Digital Library.

ECS Transactions— (ECST) is the online database containing full-text content of proceedings from ECS meetings and ECS-sponsored meetings. ECST is a high-quality venue for authors and an excellent resource for researchers. The papers appearing in ECST are reviewed to ensure that submissions meet generally-accepted scientific standards. Each meeting is represented by a volume and each symposium by an issue.

Monograph Volumes — The Society sponsors the publication of hardbound monograph volumes, which provide authoritative accounts of specific topics in electrochemistry, solid-state science, and related disciplines.

For more information on these and other Society activities, visit the ECS website:

www.electrochem.org

Chapter 1

Semiconductor Growth & Processing

ECS Transactions, 69 (14) 3-15 (2015)
10.1149/06914.0003ecst ©The Electrochemical Society

CMOS Compatible *In-situ* N-type Doping of Ge Using New Generation Doping Agents P(MH$_3$)$_3$ and As(MH$_3$)$_3$ (M=Si, Ge)

Chi Xu[a], J. D. Gallagher[a], C. L. Senaratne[b], P. E. Sims[b], J. Kouvetakis[b] and J. Menéndez[a]

[a]Department of Physics, Arizona State University, Tempe Arizona 85287, USA
[b]Department of Chemistry and Biochemistry, Arizona State University, Tempe Arizona 85287, USA

New *in-situ* methods have been developed to dope Ge-on-Si films with P and As using reactions of novel compounds P(MH$_3$)$_3$ and As(MH$_3$)$_3$ (M=Si, Ge) and UHV-CVD and molecular source epitaxy methods. In these experiments Ge$_3$H$_8$, Ge$_4$H$_{10}$ hydrides were used to deposit intrinsic Ge layers on Si(100) to produce *i*-Ge/Si(100) platforms upon which the *n*-Ge layers were grown at ultra-low temperatures of 300-330°C. The resultant *n*-Ge/*i*-Ge/Si(100) samples exhibited device quality crystallinity and defect-free microstructures as evidenced by XTEM. Infrared spectroscopic ellipsometry (IRSE) and Hall revealed carrier concentrations, mobilities and resistivities that are on par with values observed in bulk Ge. SIMS gave flat dopant profiles and abrupt transitions at the *n/i* interfaces. The highest carrier concentration was achieved in As doped samples at 8.44×10^{19} cm^{-3} and the lowest resistivity was observed in P doped samples at 4.2×10^{-4} Ω·cm. Comparison between IRSE and SIMS data revealed near full activation (>80%) of the absolute dopant concentrations in the range of 1×10^{20} cm^{-3}. The methods reported here have the potential for applications in group-IV semiconductor technologies that require high doping levels, low resistivities and shallow junction depths.

Introduction

Doping germanium has been an area of interest at the academic and industrial sectors for several decades, and it has gained extensive popularity with realizing the potential of Ge as a high performance semiconductor in modern technologies such as photovoltaics, IR photonics and microelectronics. Ge-based MISFETs are believed to be advantageous due to their higher carrier mobilities relative to current technologies. However, compared to *p*-MISFETs, the realization of Ge *n*-MISFETs analogues is facing more difficulties, because it requires ultra-high doping levels approaching 1×10^{20} cm^{-3} to reduce the high specific contact resistivity (ρ_c) caused by a higher Schottky barrier height of metal/*n*-Ge interfaces (1).

Ion implantation is probably the most thoroughly studied method for doping Ge in the past 30 years, but it has limited chance of application in the field of Ge MISFETs, for the following reasons. First, ion implantation doping requires regrowth annealing to recrystallize the Ge host lattice and to activate dopants under high temperatures which may be incompatible with for CMOS processing conditions. Second, it is almost impossible to control the doping level and depth independently, because diffusion of dopants is inevitable under high regrowth temperatures, which results in loss of dopants, lower carrier concentrations and extended junction depths (2,3,4,5).

As an emerging technique, *in-situ* doping of Ge is gaining ground over the past few years, for it has proved to be able to deliver high doping levels at low growth temperatures, and to allow control of doping level and doping depth independently, in accord with expectations for CMOS processing. The first generation of gaseous precursors used to produce n-type Ge films included GeH_4 as the Ge source, and PH_3 as the source of P dopant atoms (6,7,8). Those exploratory studies succeeded in achieving relatively high doping levels and flat dopant profiles, but still lack flexibility and practicality from a commercial device fabrication perspective. This is possibly due to the limited reactivity of these precursors at low growth temperature conditions required for some applications. Most recently new reactants that may be better suited for low-temperature processing were introduced, including Ge_3H_8 and Ge_4H_{10} as Ge sources, and $P(MH_3)_3$ and $As(MH_3)_3$ (M=Si, Ge) as *n*-type dopants. The former higher-order germanes have been applied to efficiently produce device quality Ge on Si films under CMOS compatible conditions (9,10), and preliminary studies on doping GeSn with the P and As compounds showed positive results (11). Recently a systematic study of doping Ge-on-Si films with $P(MH_3)_3$ and $As(MH_3)_3$ (M=Si, Ge) compounds has been carried out using-Ge buffered Si(100) substrates. Major achievements of these studies include: high carrier concentrations approaching 8.5×10^{19} cm^{-3}; record high electron mobilities, ultra-low resistivities and flat doping profiles with controllable depths and well-defined *n/i* interfaces. Furthermore the films were found to exhibit defect-free microsctructures as evidenced by TEM images. Finally proof-of-concept application of the technology in prototype devices was demonstrated through fabrication of *p-i-n* photodiodes with performance comparable to state-of-the-art.

Experiments

(a) Growth of the Ge/Si(100) substrates: The growths of Ge buffers on Si substrates were carried out in a gas-source molecular epitaxy (GSME) chamber. A 4" *p*-type or intrinsic Si(100) wafer was dipped in 5% HF/methanol solution for 2 minutes to hydrogen passivate the surface. It was then dried under a stream of N_2, loaded into the chamber and then out-gassed until the chamber pressure is restored at background levels of ~ 3-5×10^{-10} Torr. Several 1-min flashing cycles at 900°C were then performed to desorb the H_2 passivation and remove residual contaminants form the wafer surface. The clean surface morphology was routinely checked by RHEED. The growth of the Ge buffer was conducted at wafer temperature of approximately 350°C by introducing a mixture of 4.5 Liter-Torr Ge_4H_{10} and 90 Liter-Torr H_2 into the chamber via a needle valve. The typical chamber pressure under these conditions was kept at 1.1×10^{-4} Torr.

The growth period lasted for~50 minutes yielding Ge layers with a thickness of 1 micron. The Ge buffer was subsequently subjected to *in-situ* annealing at 650°C for 3 minutes to reduce the number of threading defects.

(a) Growth of the n-Ge/Ge/Si(100) substrates The P-doped Ge layers were categorized into two groups based on the reaction technique (GSME or UHV-CVD) used for their deposition. The first group was grown in the same GSME reactor used for the fabrication of the Ge buffers. The *n*-layer was grown immediately thereafter *in situ* directly on top of the freshly prepared intrinsic buffers. For the doping experiments mixtures of ~1 Liter-Torr Ge_4H_{10}, 20 Liter-Torr H_2 and appropriate amount of $P(MH_3)_3$ (M=Si, Ge) were admitted into the chamber, at pressures of 1.1×10^{-4} Torr. The growth temperature of the doped layers was kept at 330°C which is slightly lower than that of the Ge buffers. Typical growth times were 10-13 minutes, creating *n*-type Ge films with thicknesses of 200 nm. The activated doping levels of these samples were found to approach 3.5×10^{19} cm^{-3}. In addition silicon was incorporated in a more efficient manner, yielding Si concentrations in the layers ranging from 1.5×10^{19} cm^{-3} to 9.9×10^{20} cm^{-3}.

The second group of the *n*-type samples was produced using an ultra-high vacuum chemical vapor deposition (UHV-CVD) system in an effort to further increase the doping concertation threshold. The reactor arrangement in this case allows further activation of the chemicals by preheating them before reaching the reaction zone thereby enabling higher incorporation of the dopant species into the films. The substrates for these experiments were produced by cleaving the 4" Ge buffers into 45×45 mm quadrants to accommodate the CVD chamber dimensions. The samples were then dipped in 5% HF/H_2O solution to remove the surface oxide layer, and then loaded into the chamber, which was kept at the growth temperature of 340°C. Freshly prepared gaseous mixture of 15 Liter-Torr Ge_3H_8, 750 Liter-Torr H_2 and appropriate amount of $P(MH_3)_3$ (M=Si, Ge) were allowed into the chamber along with a constant background flow of ultra-high purity H_2 which further dilutes the co-reactants in the reaction zone. The chamber pressure was maintained at 0.200 Torr during the growth. This method in this case also targeted *n*-type Ge layers with thicknesses of ~ 200 nm. These were found to contain active carrier doping levels up to 6.2×10^{19} cm^{-3}. The Si content in this cases was found to be at doping levels nearly equivalent to those of P indicating the $P(SH_3)_3$ precursor eliminates the SiH_3 ligands as SiH_4 products that are pumped away and do not participate in the deposition process under the low temperature reaction conditions employed.

The depositions of As-doped samples were carried out in the UHV-CVD reactor, using mixtures of 15 Liter-Torr Ge_3H_8, 750 Liter-Torr H_2 and appropriate amounts of the $As(MH_3)_3$ (M=Si, Ge) precursors. The growth temperature in this case was 330°C and the pressure was 0.200 Torr yielding active carrier densities of 8.44×10^{19} cm^{-3} which are higher than those obtained using the $P(MH_3)_3$ precursors. It is worth noting that attempts to grow As doped Ge films in the GSME reactor did not produce any viable samples with reasonable thicknesses and compositions in contrast to the P analogs. This is likely due to the lower thermal stability of the As compound which under the low pressure conditions (10^{-4} Torr) of the technique likely eliminates AsH_3 byproducts at the growth front which do not actively partake in the growth process. In the case of the CVD depositions the much higher pressures (0.200 Torr) employed in combination with the thermal pre-reactivation of the precursors in the hot walls of the chamber generate reactive gas phase intermediates which then promote facile incorporation of Ge-As bonding units into the films. The CVD grown samples were subsequently characterized by a range of methods to determine their structural, morphological and electronic properties. Infrared spectroscopic ellipsometry (IRSE) was used to extract layer thicknesses, active carrier concentrations, layer resistivity ρ, relaxation time τ and mobility μ. Hall measurements yielded carrier concentrations, resistivity ρ and mobility μ values which were in good agreement with the IRSE counterparts. XRD measurements showed that the films were mono-crystalline materials perfectly aligned with the substrates. This data were used to measure the lattice parameters, residual crystal strains and Si contents in certain cases where the content of atoms in the layers was sufficiently high within the sensitivity limit of the technique. RBS channeling corroborated the excellent level of crystallinity and epitaxial registry at the interfaces. The random spectra were used to confirm the atomic compositions and the layer thicknesses. XTEM images showed epitaxial growth of low defectivity n-layers on top of Ge buffers. SIMS measurements were used to estimate the chemical concentrations using carefully calibrated standard and the measured elemental profiles. The SIMS analyses generated useful information regarding dopant distributions and activation ratios.

Results and Discussions

Table I summarizes representative samples of P doped Ge, showing doping reagents, growth technique, along with Si and P concentrations obtained from SIMS measurements, carrier concentration N, resistivity ρ, relaxation time τ and mobility μ values obtained from IRSE. Similar growth parameters and electrical results of selected As doped Ge samples are shown in Table II. An early account of structural and electrical properties of these materials could be found in ref. (12,13).

TABLE I. Summary of P doped Ge samples.

Precursor	Reactor	$P(MH_3)_3$ L-Torr	SIMS Si (cm^{-3})	SIMS P (cm^{-3})	Carrier conc. IRSE (cm^{-3})	Resistivity ρ ($\Omega \cdot cm$)	Relaxation time τ $(10^{-15}\ s)$	Mobility μ (cm^2/Vs)
$P(SiH_3)_3$	GSME	0.01	1.5E19	3.5E18	4.0E18	0.0028	38	557
$P(SiH_3)_3$	GSME	0.02	3.4E19	6.5E18	7.7E18	0.00165	36.5	535
$P(SiH_3)_3$	GSME	0.04	4.0E20	4.3E19	3.1E19	0.00066	20.5	300
$P(SiH_3)_3$	GSME	0.06	5.3E20	4.7E19	3.5E19	0.00066	18.5	271
$P(SiH_3)_3$	GSME	0.10	9.9E20	6.1E19	3.3E19	0.0006	21.8	320
$P(GeH_3)_3$	GSME	0.003	--	3.5E18	4.2E18	0.00207	50	733
$P(GeH_3)_3$	GSME	0.03	--	1.9E19	1.8E19	0.00075	31.2	457
$P(GeH_3)_3$	GSME	0.06	--	2.7E19	2.8E19	0.00067	23	337
$P(GeH_3)_3$	GSME	0.10	--	4.8E19	3.0E19	0.00056	25.5	374
$P(GeH_3)_3$	GSME	0.15	--	8.1E19	2.8E19	0.00058	26.5	388
$P(SiH_3)_3$	CVD	0.15	2.2E19	5.6E19	4.2E19	0.00052	19.5	286
$P(SiH_3)_3$	CVD	0.30	6.2E19	7.1E19	5.7E19	0.00047	16	235
$P(GeH_3)_3$	CVD	0.20	--	7.0E19	6.0E19	0.000415	17	249
$P(GeH_3)_3$	CVD	0.30	--	1.3E20	6.2E19	0.00042	16.3	239

TABLE II. Summary of As doped Ge samples.

Precursor	As(MH$_3$)$_3$ L-Torr	SIMS Si (cm^{-3})	SIMS As (cm^{-3})	Carrier concentration IRSE (cm^{-3})	Resistivity ρ ($\Omega\cdot$cm)	Relaxation time τ (10^{-15} s)	Mobility μ (cm^2/Vs)
As(SiH$_3$)$_3$	0.04	3.3E18	9.0E18	8.9E18	0.0018	26.5	388
As(SiH$_3$)$_3$	0.04	8.0E18	1.45E19	1.69E19	0.00122	20.7	303
As(SiH$_3$)$_3$	0.08	1.78E18	4.80E19	3.99E19	0.00068	15.7	230
As(SiH$_3$)$_3$	0.15	6.59E19	9.09E19	7.31E19	0.000525	11.1	163
As(SiH$_3$)$_3$	0.30	5.68E19	7.80E19	6.09E19	0.00053	13.2	193
As(SiH$_3$)$_3$	0.38	7.25E19	1.04E20	8.30E19	0.000513	10	147
As(SiH$_3$)$_3$	0.45	8.90E19	1.30E20	7.47E19	0.0006	9.5	139
As(GeH$_3$)$_3$	0.15	--	2.52E19	2.48E19	0.0010	17.2	252
As(GeH$_3$)$_3$	0.30	--	9.06E19	7.83E19	0.00049	11.1	163
As(GeH$_3$)$_3$	0.38	--	1.30E20	8.44E19	0.00049	10.3	151
As(GeH$_3$)$_3$	0.45	--	2.14E20	6.82E19	0.00052	12	176

XRD measurements of P(SiH$_3$)$_3$ doped samples grown by GSME show that the n-layer exhibits 004 peaks that shift to lower Bragg angles with increasing P(SiH$_3$)$_3$ concentration in the reaction mixture, indicating the incorporation of significant amounts of substitutional Si in the doped Ge layers, as shown in Figure 1. Estimates of the Si content from relaxed lattice constants and of RBS spectra indicate that the amounts vary up to atomic 3% with increasing active layer contents up to ~6×10^{19}/cm^3. The Si:P ratios in these samples are considerably higher than the expected 3:1 value on the basis of intact incorporation of the Si3P molecular core of the precursor. In contrast, the P(SiH$_3$)$_3$ doped Ge samples grown in UHV-CVD showed no extra peaks in XRD spectra (the 004 peaks of the Ge buffer and Ge doped layer overlap indicating identical lattice dimensions) in spite of their higher doping concentrations compared to GSME samples. The Si:P ratios in this case are consistently in the 1:1 range, much lower than the expected 3:1 ratio, corroborating the fact that different growth mechanism are in operation between the two techniques.

For the As(SiH$_3$)$_3$ doped Ge samples, the amount of silicon is invariably lower in the 10^{19}/cm^3 range irrespective of As content and growth conditions indicating that the silicon substitution is kept at impurity levels. For samples with As carrier densities above 4×10^{19}/cm^3 the XRD spectra show a shoulder on the left side of the 004 buffer layer peak corresponding to the doped epilayer indicating a slightly larger lattice constant attribute to the alloying with As. The observed lattice expansion agrees with previously reported theoretical results which indicate that electronic and size effects are the reason for this manifestation. (14).

XTEM studies were conducted to investigate the crystallinity and defect densities of the n-type and intrinsic Ge layers. Figure 2 shows a representative TEM image for a P(SiH$_3$)$_3$ doped sample grown in CVD, with a carrier concentration of 5.7×10^{19} cm^{-3}. Occasional defects could be seen at the lower portion near the Ge/Si interface, and the majority of i-Ge buffer and the whole n-Ge layer are defect free. The top surface and the n-i interface are both flat, indicating that high quality epitaxial growth has been realized at such high doping levels.

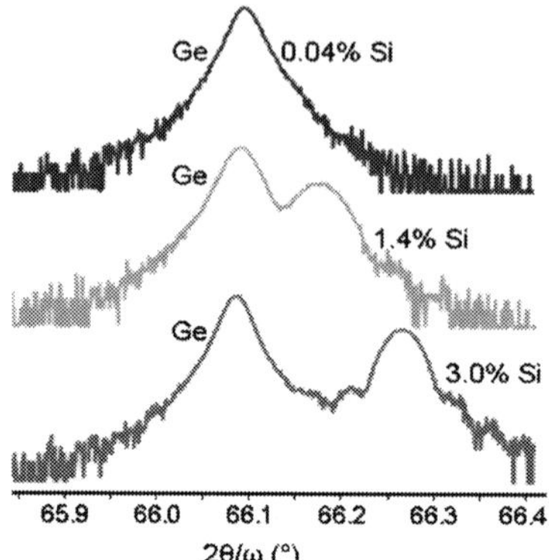

Figure 1. XRD 004 spectra of representative $P(SiH_3)_3$ doped Ge samples grown in GSME on top of Ge buffered Si. The top sample contains 0.04% Si in the *n*-layer, whose XRD peak overlapped with the *i*-Ge buffer peak. The lower samples contain more Si in the *n*-layers, and separate GeSi(P) peaks could be seen at higher Bragg angles. The active carrier concentrations for these samples are $4\times10^{18}/cm^3$, $3.1\times10^{19}/cm^3$ and $3.3\times10^{19}/cm^3$, respectively.

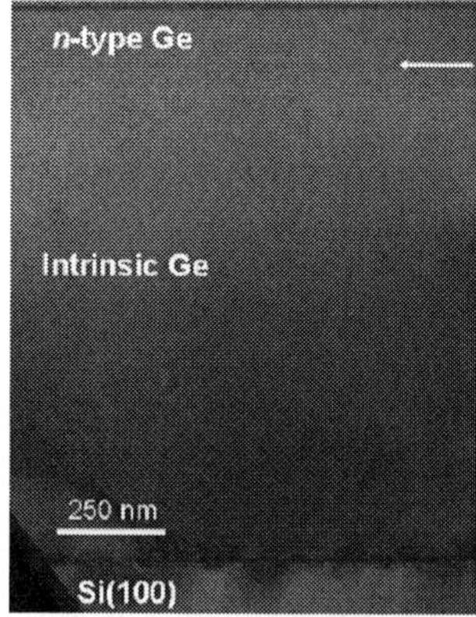

Figure 2. XTEM image of a sample doped with P(SiH3)3 in CVD, which has an active carrier concentration of 5.7×10^{19} cm^{-3}, showing low defect density, flat surface and uniform *n-i* interface (marked by arrow).

AFM images were obtained to examine the surface morphology of the doped Ge layers. Smooth surfaces with RMS roughness values below 1 nm were commonly observed, indicating that the growth of doped Ge followed a layer-by-layer mechanism. Figure 3 below shows a representative AFM image for an $As(GeH_3)_3$ doped sample with a carrier concentration of 8.44×10^{19} cm^{-3}, which is the highest in this study. The RMS roughness for a 20×20 μm scan from this sample is 0.69 nm.

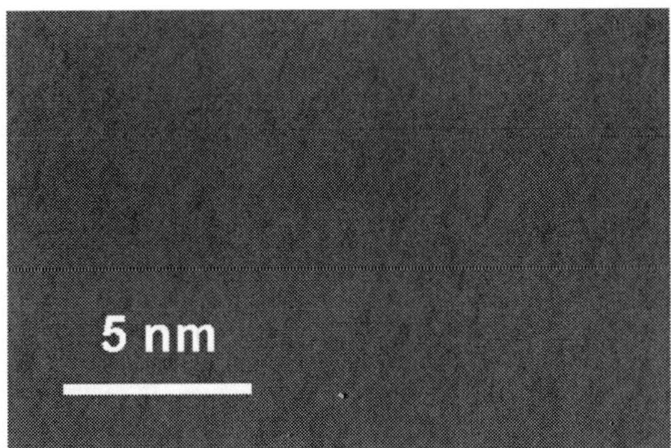

Figure 3. AFM image of an As(GeH$_3$)$_3$ doped sample with N=8.44×10^{19} cm^{-3}.

SIMS measurements were performed for all P(MH$_3$)$_3$ and As(MH$_3$)$_3$ samples using a Cs$^+$ source to determine the atomic distributions in doped layers. Uniform P/As profiles in the n-layers and abrupt transitions at the n-i interfaces were observed in all cases irrespective of dopant concertation. The absolute dopant amounts determined by these measurements were compared with the active carrier densities obtained by IRSE and Hall and the results showed near full activation of the P and As atoms, even at doping levels far beyond the P or As solubility limits. This is illustrated in Figure 4 which shows a typical SIMS profile of a sample doped with P(GeH$_3$)$_3$ using the CVD technique. The plots in the figure show a flat P profile corresponding to total P concentration of 7.0×10^{19} cm^{-3}. The IRSE gave a carrier concentration of 6.0×10^{19} cm^{-3}, indicating excellent carrier activation ratio as expected.

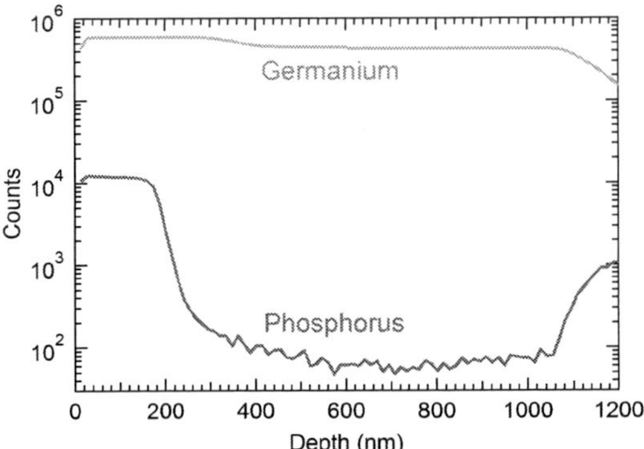

Figure 4. SIMS profile of n-Ge/Ge/Si(100) sample doped with P(GeH$_3$)$_3$ using CVD. The P content within the n-layer is 7.0×10^{19} cm^{-3} and drops precipitously at the interface with the intrinsic Ge buffer. The flat atomic profile across the entire structure observed here is typical for all samples produced in this study.

Hall measurements to determine the carrier densities have been carried out on all samples in this study and good agreements with ellipsometry have been found.

Nonetheless, small mismatches between the results from these two methods still exist, which appear to be precursor dependent. The P doped samples on average registered Hall carrier concentrations higher by 13% than their respective IRSE values. The As(SiH$_3$)$_3$ doped samples showed Hall carrier concentrations on average 15% lower than IRSE values, whereas the As(GeH$_3$)$_3$ doped samples exhibited almost perfect agreement between Hall and ellipsometry data. The fundamental cause for such systematic differences is out of scope of this paper, but it could be related to alloying effects as well as Hall factor issues.

The incorporation ratios (i.e. the total amount of dopant atoms in the solid film vs the amount of dopant atoms in the mixture for a given sample) are slightly different for CVD and GSME samples, as shown in Figure 5. The horizontal axis indicates the atomic concentration of dopants in the gaseous mixtures, and the vertical axis shows corresponding total concentrations of P or As atoms incorporated into the layers. Both axes present percentage values of the data. We note that both GSME and CVD growths follow linear trends. The CVD method exhibits a higher efficiency of atomic incorporation as indicted by the slope of the line which is almost 6 times steeper than that of GSME. This observation could be partially explained by the reactor configuration which in the CVD case the gaseous reactant mixtures are thermally pre-activated before reaching the substrate surface, by flowing through the hot chamber for a distance of at least 20 cm. In contrast, the GSME process takes place in a cold wall single wafer system where thermal pre-activation is minimal.

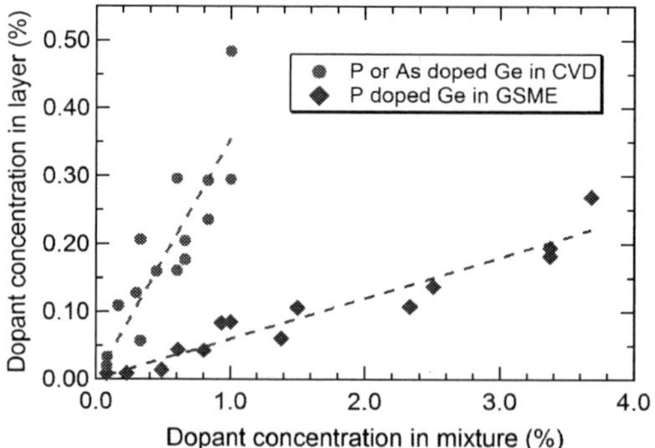

Figure 5. Relationships between total dopant concentrations determined from SIMS with the corresponding concentrations in gaseous mixtures. The red data points correspond to depositions of P(MH$_3$)$_3$ and As(MH$_3$)$_3$ in CVD. The blue data point correspond to samples produced via P(MH$_3$)$_3$ in GSME.

Figure 6 plots active carrier concentrations versus corresponding dopant atom concentrations in the gaseous reaction mixtures (the number of P/As dopant atoms in the gaseous mixture over total number of atoms including Ge in the mixture). The activation of P dopants is different in the two techniques as shown by the green and blue data points in the plot. At low mixture ratios the active carriers increase with increasing gaseous dopant concentrations up to an upper limit, beyond which addition of more dopants atoms tends to decrease the active carriers in the films. In the CVD experiments the saturation of P active carriers occurs at ~0.7% dopant concertation, which is smaller than the 2.5% limit observed in the GSME experiments. In the case of As the trend is similar to that of P except the active carrier concertation is higher than that of P for a give atomic mixture. All three sets of data presented in the plots of Figure 6 could be fitted by second order polynomials, as shown by dashed lines in the figure.

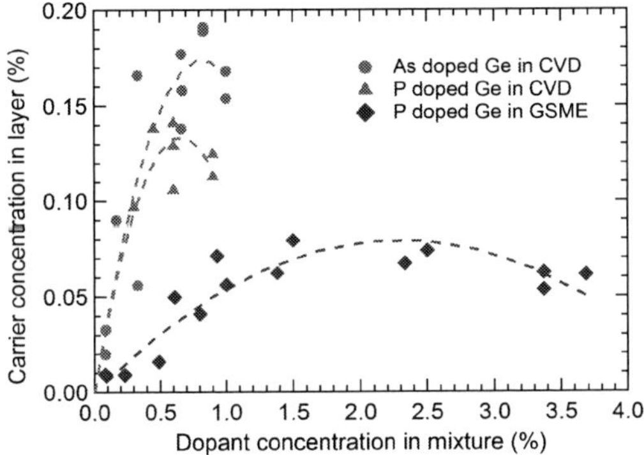

Figure 6. Trends between active carrier concentrations in doped Ge layers and the corresponding P/As atom concentrations in gaseous mixtures used for depositions of the precursors using CVD and GSME techniques.

A combined analysis of the data in figure 5 and 6 provides insights with regards the activation ratios of the samples produced using CVD and GSME depositions of all precursors. The activation ratio is defined as the ratio between active carrier concentrations obtained from IRSE and the total dopant content obtained from SIMS for a given sample. Figure 7 plots the activation ratios of P for GSME samples (blue squares) and P/As for CVD samples (red dots). Dashed curves are empirical trend lines. Both systems yield near 100% activation at lower total dopant concentrations (SIMS values) below the range of 3-4×10^{19} cm^{-3}. The activation gradually decreases until saturating near 20-30% activation levels. The CVD system shows a less steep slope in the trend line than GSME indicating that the former technique is capable of achieving ~80% activation of the dopant atoms up to 1×10^{20} cm^{-3}. This indicates a higher potential for the latter in applications requiring hyper doping Ge at levels approaching 10^{20} cm^{-3}. In contrast, the activation ratio for GSME at 1×10^{20} cm^{-3} total dopant concentrations is already below 30%.

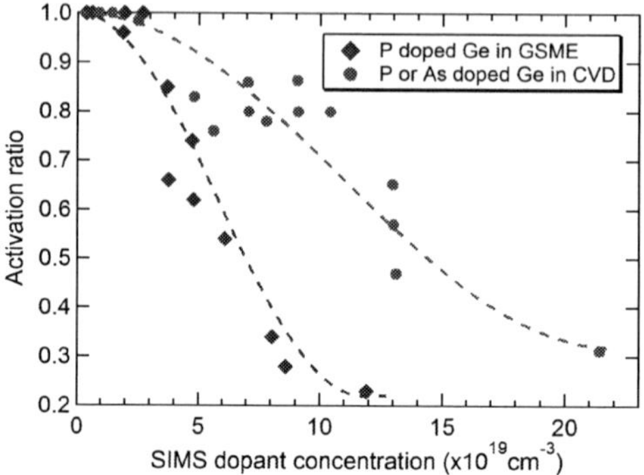

Figure 7. Plot of activation ratios vs SIMS total dopant concentrations for CVD and GSME samples.

Mobilities for P and As doped samples obtained from CVD and GSME depositions of all precursors are plotted in Figure 8 where they are compared to published results for n-type bulk Ge samples doped with P, As and Sb. The upper black line is a fit to bulk Sb doped Ge data in ref (15), and the lower blue line is a fit to bulk As doped Ge data from ref (16).The grey band between the lines is the region of mobilities expected from As, P and Sb doped bulk Ge samples. The red and orange dots are IRSE mobility data from As doped samples produced using $As(MH_3)_3$ precursors in this study. They follow the mobility data for bulk Ge perfectly without indication of any difference between the Si- and Ge-compounds. The black points are data from P doped samples produced using $P(MH_3)_3$ in this work. They agree with corresponding mobilities of P doped bulk Ge. Note that the mobilities of P doped samples are higher than the As analogs as expected. This observation agrees with theoretical calculations reported in Ref. 17 which computed central-cell corrections to the Coulomb scattering potentials for different kind of impurities and predicted a mobility order of Sb>P>As in *n*-type doped Ge.

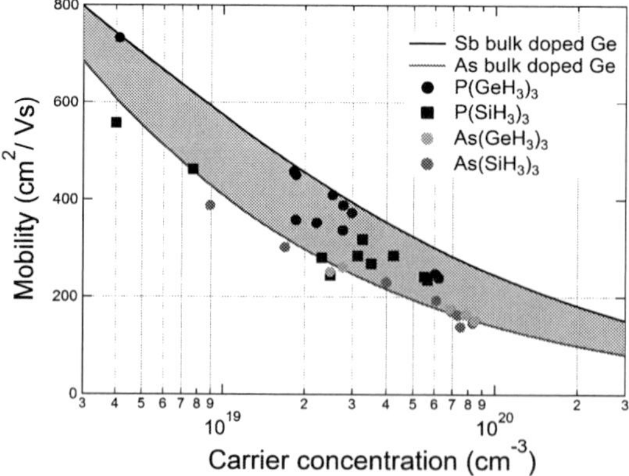

Figure 8. Plot of mobilities for P and As doped samples. Both data sets mimic the values of bulk doped Ge analogs.

Figure 9 compares the growth rates for the sets of silyl- [P(SiH$_3$)$_3$, As(SiH$_3$)$_3$] and germyl- [P(GeH$_3$)$_3$, As(GeH$_3$)$_3$] compounds as a function of gas mixture dopant concertation in the deposition reactions. For a given gas dopant concertation no significant difference in growth rate is seen within each group IV set of precursors. For the silicon-based compounds, the growth rates remain constant at ~ 3.5nm/min under our growth conditions over the entire range of concentrations studied in this work. The growth rates of the germanium-based compounds are similar to those of the silicon counterparts at low gas dopant concertation below 0.6%, however a marked increase in growth rate could be seen with increasing dopant concentration beyond 0.6% for the Ge compounds (red dots).

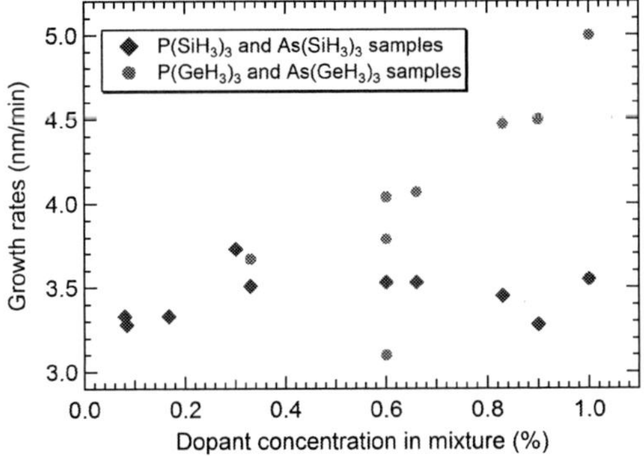

Figure 9. Comparison of growth rates from silyl- and germyl-compounds. The Si compounds showed no effect on growth rates, while the Ge compounds show higher rates with increasing gas phase dopant concentrations.

Room temperature vapor pressures for these four doping agents are listed in Table III. The silyl-compounds have much higher vapor pressures than their germyl-counterparts. In our hands, the Si compounds could be stored at room temperature for months without noticeable decomposition, under a blanket of H$_2$ ambient. In contrast, the Ge compounds need to be stored in a refrigerator in order to prevent slow decomposition. Given the fact that silyl-and germyl-reagents have showed similar doping capabilities, and that the Si incorporation in growths from silyl-compounds is usually low using CVD, it could be concluded that P(SiH$_3$)$_3$ and As(SiH$_3$)$_3$ should be more practical for industrial scale-up purposes.

TABLE III. Vapor pressures for different doping reagents used in this study.

Compounds	P(SiH$_3$)$_3$	P(GeH$_3$)$_3$	As(SiH$_3$)$_3$	As(GeH$_3$)$_3$
Vapor pressure at 20°C (Torr)	22	1.5	10	0.6

Summary

Single-source doping sources $P(MH_3)_3$ and $As(MH_3)_3$ (M=Si, Ge) were used to produce n-type Ge films grown on i-Ge/Si(100) platforms. Highly-doped Ge layers were produced, exhibiting device quality crystallinity, low defect densities and flat dopant profiles with well-defined n-i interfaces. Comparison between IRSE, Hall and SIMS data revealed donor activation ratios above 80% of the absolute chemical concentrations in the range of 1.0×10^{20} cm^{-3}. The $P(MH_3)_3$ compounds were utilized to conduct GSME and UHV-CVD depositions in order to investigate the growth mechanism under a wide range of pressure, temperature and reaction flux conditions. The highest doped films in this case were achieved from the CVD experiments at 6.2×10^{19} cm^{-3} showing resistivities of 4.2×10^{-4} $\Omega \cdot$cm. The As compounds were only studied using CVD depositions, yielding doping levels of 8.4×10^{19} cm^{-3} and resistivities of 4.9×10^{-4} $\Omega \cdot$cm. The As sample mobilities were lower than those of the P analogs, consistent with theoretical predictions. In both cases the mobilities were on par with those reported for bulk Ge. The Si incorporation in GSME experiments using $P(SiH_3)_3$ exceeded the expected 3:1 ratio of the molecular core, whereas in the case of the CVD experiments the ratio of incorporated Si and P was well below the 3:1 expectation, indicating differences in growth mechanisms for the two techniques. From a practical perspective, the $P(SiH_3)_3$ and $As(SiH_3)_3$ compounds might be more attractive than the $P(GeH_3)_3$ and $As(GeH_3)_3$ due to their significantly higher vapor pressures by at least 10 times. Furthermore, $P(SiH_3)_3$ and $As(SiH_3)_3$ only incorporate impurity levels of Si (<0.2%) into the films under CVD conditions and these low levels do not affect the electrical and optical properties of the doped layers in any meaningful manner. The results reported in this paper indicate potential for applications in Ge on Si based technologies that require ultra-high doping and low resistivities under low temperature growth conditions, such as group-IV lasers and Ge MISFETs.

Acknowledgement

We gratefully acknowledge the use of facilities within the Center for Solid State Science at Arizona State University. This work was supported by the Air Force Office of Scientific Research under contracts DOD AFOSR FA9550-12-1-0208 and DOD AFOSR FA9550-13-1-0022, and the National Science Foundation DMR-1309090.

References

1 . Y. Moriyama, Y. Kamimuta, Y. Kamata, K. Ikeda, A. Sakai and T. Tezuka, *Appl. Phys. Express* **7**, 106501 (2014)

2 . P. Tsouroutas, D. Tsoukalas, I. Zergioti, N. Cherkashin and A. Claverie, *J. Appl. Phys.* **105**, 094910 (2009)

3 . E. Simoen, A. Satta, A. D'Amore, T. Janssens, T. Clarysse, K. Martens, B. De Jaeger, A. Benedetti, I. Hoflijk, B. Brijs, M. Meuris and W. Vandervorst, *Mater. Sci. Semicond. Process.* **9**, 634 (2006)

4 . G. D. Dilliway, R. van den Boom, B. van Daele, F.E. Leys, T. Clarysse, B. Parmentier, A. Moussa, C. Defranoux, A. Benedetti, O. Richard, H. Bender, E. Simoen and M. Meuris, *ECS Trans.* **3**, 599 (2006)

5. C. H. Poon, L. S. Tan, B. J. Cho and A. Y. Du, *J. Electrochem. Soc.* 152 G895 (2005)

6 . S. Jang, K. Liao and R. Reif, *J. Electrochem. Soc.* **142**, 3520 (1995)

7 . J. Liu, X. Sun, R. Camacho-Aguilera, L. C. Kimerling and J. Michel, *Opt. Lett.* **35**, 679 (2010)

8 . E. Tutuc, J. Chu, J. Ott and S. Guha, *Appl. Phys. Lett.* **89**, 263101 (2006)

9 . G. Grzybowski, L. Jiang, R. T. Beeler, T. Watkins, A. V. G. Chizmeshya, C. Xu, J. Menéndez and J. Kouvetakis, *Chem. Mater.* **24**, 1619 (2012)

10. C. Xu, R. T. Beeler, L. Jiang, G. Grzybowski, A. V. G. Chizmeshya, J. Menéndez and J. Kouvetakis, *Semicond. Sci. Tech.* **28**, 105001 (2013)

11. J. Xie, J. Tolle, V.R. D'Costa, C. Weng, A.V.G. Chizmeshya, J. Menéndez and J. Kouvetakis, *Solid State Electron.* **53**, 816–823 (2009)

12. C. Xu, J. D. Gallagher, P. Sims, D. J. Smith, J. Menéndez and J. Kouvetakis, *Semicond. Sci. Technol.* **30**, 045007 (2015)

13. C. Xu, C. L. Senaratne, J. Kouvetakis and J. Menéndez, *Appl. Phys. Lett.* **105**, 232103 (2014)

14. A. Chizmeshya, C. Ritter, J. Tolle, C. Cook, J. Menéndez and J. Kouvetakis, *Chem. Mater.* **18**, 6266 (2006)

15. D. B. Cuttriss, *Bell Syst. Tech. J.* **40**(2), 509 (1961)

16. W. G. Spitzer, F. A. Trumbore, and R. A. Logan, *J. Appl. Phys.* **32**(10), 1822 (1961)

17. H. I. Ralph, G. Simpson and R. J. Elliot, *Phys. Rev. B.* **11**(8), 2948-2956 (1975)

ECS Transactions, **69** (14) 17-32 (2015)
10.1149/06914.0017ecst ©The Electrochemical Society

Propagation of Nanopores and Formation of Nanoporous Domains during Anodization of n-InP in KOH

D. Noel Buckley[a], Robert P. Lynch[a], Nathan Quill[a], and Colm O'Dwyer[b]

[a] Department of Physics & Energy, and Materials & Surface Science Institute, University of Limerick, Ireland

[b] Department of Chemistry, and Tyndall National Institute, University College Cork, Ireland

Anodization of highly doped (10^{18} cm^{-3}) n-InP in $2 - 5$ mol dm^{-3} KOH under potentiostatic or potentiodynamic conditions results in the formation of a nanoporous sub-surface region. Pores originate from surface pits and an individual, isolated porous domain is formed beneath each pit in the early stages of anodization. Each such domain is separated from the surface by a thin non-porous layer (typically ~40 nm) and is connected to the electrolyte by its pit. Pores emanate from these points along the <111>A crystallographic directions to form domains with the shape of a tetrahedron truncated symmetrically through its center by a plane parallel to the surface of the electrode. We propose a three-step model of electrochemical pore formation: (1) hole generation at pore tips, (2) hole diffusion and (3) electrochemical oxidation of the semiconductor to form etch products. Step 1 determines the overall etch rate. However, if the kinetics of Step 3 are slow relative to Step 2, then etching can occur at preferred crystallographic sites leading to pore propagation in preferential directions.

INTRODUCTION

In many materials, anodic etching leads to the formation of porous structures. In metals, localized etching may be promoted by the formation of semi-conducting, metal-oxide/hydroxide layers which, due to interlayer forces, may be molded into tubular structures. In semiconductors, localized etching can occur leading to selective removal of material such that the remaining material forms a skeletal structure that encompasses a network of pores. Si, SiC, Ge, Ge/Si alloys, and various II-VI and III-V compounds (including InP) can be made porous in such a manner[1-13].

The morphologies of these pores vary in orientation, frequency of branching, type of infilling and extent of the porous structure. For instance, porous layers form in GaP anodized in aqueous H_2SO_4 solution by the growth of almost hemispherical domains of pores into continuous porous layers[14]. Such domains form due to the radial propagation of their pores from pits in the electrode surface. However, pore propagation can also occur along crystallographic directions and changing the electrode potential can change the propagation direction of these pores. Similar variations in pore morphology are observed in InP and GaAs. In one of the first observations of porous InP formation, pores with triangular cross sections were reported to form along <111> directions when (11$\bar{1}$)A-oriented n-InP was anodized in the dark in aqueous HCl via an array of periodic holes in a Si mask.[15] Similar pores in both GaAs and InP anodized in HCl[16] have been observed to propagate along the <111>A directions. At higher potentials, propagation of such pores deviates from the crystallographic directions[17,18] towards the direction of the source of current[19,20] and switching of potential allows alternation between these two regimes of "crystallographically-oriented" (CO) and "current-line-oriented" (CLO) pore propagation.[21]

It is generally accepted that pore propagation in highly doped n-type semiconductors is controlled by hole generation under the influence of a high electric field due to the small radius of curvature at the pore tip. Zhang[22] modeled the relationship between pore-tip shape and electric field in silicon and showed that the electric field at the surface is sufficiently enhanced by the pore-tips curvature to enable substantial tunneling of carriers. This results in significant etching occurring only at pore-tips, allowing continued propagation of these tips into the substrate.

Both chemical and electrochemical etching[23,24] of III-V semiconductors show preferential etching of {111}B planes (*i.e.* group-V-terminated planes). The slowest-etching plane is usually {111}A and so these facets are revealed during etching of InP, GaAs and GaP. Due to the differing etch rates of crystal planes, the formation of tetrahedral etch pits (seen as dove-tailed and v-groove voids, respectively in (0$\bar{1}\bar{1}$) and (0$\bar{1}$1) cross sections) is observed on the (100) surface of III-V semiconductors.[23]

This paper reviews our work[1-3,5,17] on pore propagation and the resulting domain shape in the InP-KOH system. We propose a model[2] based on pore propagation along the <111>A directions to explain the observed structures and domain shapes, and we compare the predictions of the model quantitatively with the experimental observations. We examine why pore propagation is crystallographically oriented even though the etching is controlled by hole generation at pore tips and propose a three-step model[1] based on competitive kinetics between electrochemical reaction and hole diffusion. The model is supported by a variety of SEM and TEM observations of InP electrodes anodized under various conditions.

EXPERIMENTAL

Wafers were monocrystalline, sulfur-doped, n-type indium phosphide (n-InP) grown by the liquid-encapsulated Czochralski (LEC) method and supplied by Sumitomo Electric. They were polished on one side and had a surface orientation of (100) and a carrier concentration in the range 3–6 × 10^{18} cm^{-3}. Crystallographic orientation was indicated by primary and secondary flats marking the natural {011} cleavage planes of the wafer according the European/Japanese system. The manufacturer identified these planes from the 'dovetail' and 'V-groove' etch patterns revealed by a standard wet chemical etch. Thus, the primary flat was chosen so that the {111} plane intermediate in direction between it and the (100) surface plane is a {111}A plane , *i.e.* In terminated. For convenience we will call the plane of the primary flat an α plane, the secondary flat a β plane and the wafer surface a γ plane. The schematic in Fig. 1 summarizes the various planes and directions.

To fabricate working electrodes, wafers were cleaved into coupons (typically ~5 mm square) along the α and β planes noting their orientation. Ohmic contact was made by alloying indium to the back of a coupon; the back and the cleaved edges were then isolated from the electrolyte by means of a suitable varnish. The electrode area was typically 0.2 cm^2. Prior to immersion in the electrolyte, the working electrode was immersed in a piranha etchant (3:1:1 H_2SO_4:H_2O_2:H_2O) for 4 minutes and then rinsed with deionized water.

Anodization was carried out in 5 mol dm^{-3} aqueous KOH at room temperature in the absence of light using a linear potential sweep at 2.5 mV s^{-1}. A conventional three-electrode cell configuration was used, employing a platinum counter electrode and a saturated calomel electrode (SCE) to which all potentials are referenced. A CH Instruments Model 650A Electrochemical Workstation interfaced to a Personal Computer (PC) was employed for cell parameter control and for data acquisition.

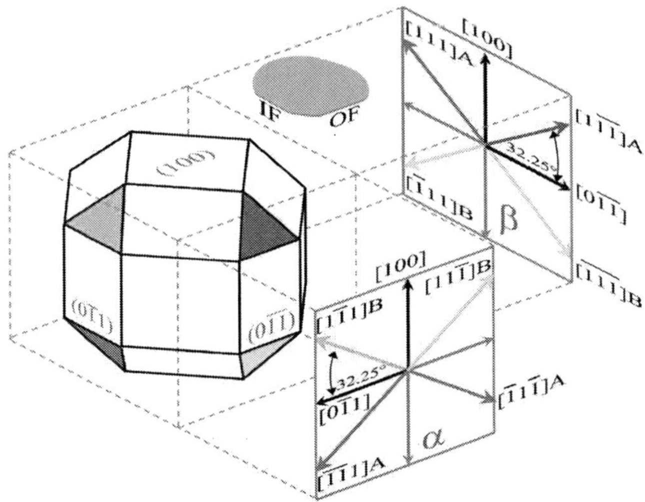

Fig. 1 Isometric drawing showing crystallographic directions and planes relative to the surface and the primary (OF) and secondary (IF) flats of a wafer. We call the primary flat (01̄1̄) an α plane and the secondary flat (01̄1) a β plane. The [11̄1̄]A direction is shown in the β plane and between the [100] and [01̄1̄] directions. This is identified as a [11̄1̄]A direction (*i.e.* In-terminated) from the wafer specification because [100] and [01̄1̄] represent the directions of the surface and primary flats, respectively. Likewise, [11̄1]B is in the α plane and between [100] and [01̄1] as shown.

Cleaved α and β cross sections of electrodes were examined using a Hitachi S-4800 field-emission scanning electron microscope (FE SEM) operating at 5 kV, unless otherwise stated. Electron-transparent sections for plan-view and cross-sectional transmission electron microscopy (TEM) examination were prepared using standard focused ion beam (FIB) milling procedures in an FEI 200 FIB workstation. The TEM characterization was performed using a JEOL (2000FX and 2011) transmission electron microscope operating at 200 kV.

RESULTS AND DISCUSSION

1. Electrochemical Growth of Nanoporous InP

Fig. 2a shows an LSV of an n-InP electrode in 5 mol dm^{-3} KOH. The potential was scanned at 2.5 mV s^{-1} from an initial potential of 0 V. There is little current at potentials less than 0.3 V, but continued anodization to potentials greater than 0.4 V results in a rapid increase in the current density to a peak value of 20 mA cm^{-2} at 0.48 V. Above 0.48 V, the current density decreases quite rapidly, reaching a value of ~3 mA cm^{-2} at 0.6 V. A significant anodic oxidation process clearly occurs above ~0.4 V and becomes self-limiting at higher potentials. Fig. 2b shows a LSV of a similar n-InP electrode in 1 mol dm^{-3} KOH at the same scan rate (2.5 mV s^{-1}). As in Fig. 2a, an anodic peak can similarly be observed, but at a higher potential (~1.25 V).

The electrodes formed during anodization in the two KOH electrolyte concentrations were examined in TEM. The micrographs shown in Fig. 3 demonstrate that the anodic processes are remarkably different in the two different concentrations of KOH. Fig. 3a shows that anodization in 5 mol dm^{-3} KOH results in the formation of a nanoporous InP layer and it should be noted that similar nanoporous layers were observed in 2 mol dm^{-3} and 3 mol dm^{-3} KOH. Fig. 3a was taken in an under-focus condition and thereby uses phase contrast effects to enhance the visibility of the pores formed within the upper part of the InP. The sense of the Fresnel contrast changes seen as a function of the defocus conditions in through focal series of bright field micrographs confirmed the formation of the low density regions described, although it should be emphasized that the nanopores are heavily interspersed with a high volume fraction of microcrystalline material. Fig. 3b, by comparison, shows that, after a similar potential sweep, there appears to be little change in the electrode during anodization in 1 mol dm^{-3} KOH other than a thin (~40 nm) surface film, as marked at A.

LSVs were performed in 5 mol dm^{-3} KOH over a range of scan rates and in all cases an anodic current peak was observed as in Fig. 2a. While the current densities increase at higher scan rates, the charge density corresponding to the anodic peak was found to remain relatively constant (0.57 C cm^{-2}), indicating that an approximately equal amount of InP is oxidized irrespective of the scan rate.

Further investigation by TEM shows the porous region is capped by a thin layer (~40 nm) close to the surface that appears to be unmodified. The thickness of the nanoporous region is generally uniform although in some areas the interface with the substrate is characteristically non-planar. The average pore width and inter-pore distances are noted to be similar (~40 nm) to one another as well as to the thickness of the near-surface layer.

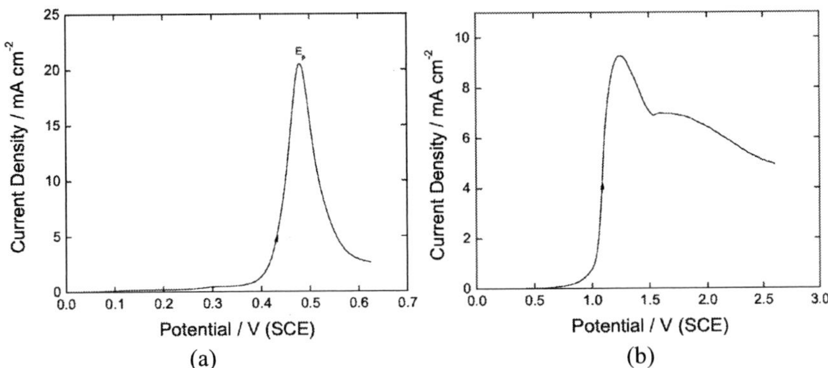

(a)   (b)

Fig. 2 LSV at 2.5 mV s^{-1} of an n-InP electrode in (a) 5 mol dm^{-3} KOH; (b) 1 mol dm^{-3} KOH.

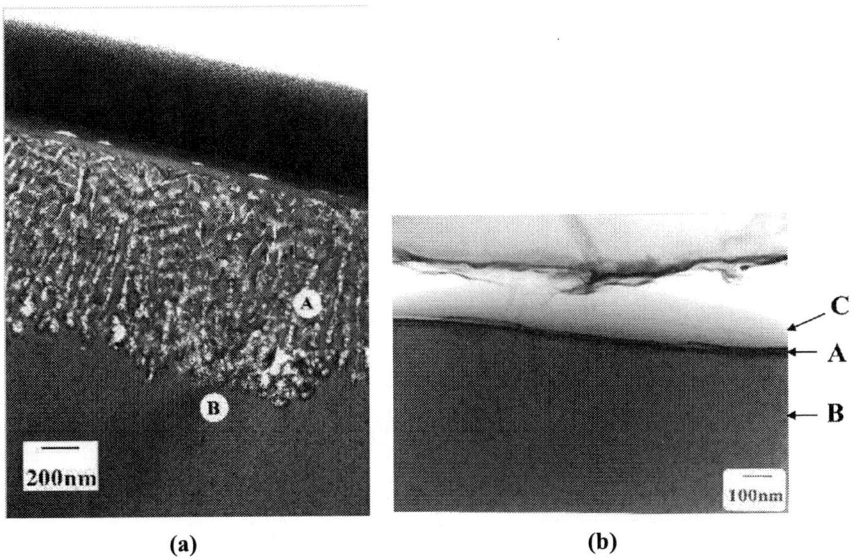

(a)   (b)

Fig. 3 (a) Under-focus bright field TEM micrograph of an n-InP electrode after an LPS from 0 V to 0.63 V (SCE) at 2.5 mV s^{-1} in 5 mol dm^{-3} KOH. A nanoporous InP layer (A) is observed above the InP substrate (B) and beneath a dense near-surface layer (C). (b) Bright field TEM micrograph of an n-InP electrode cross-section after a LPS from 0 V to 1.3 V (SCE) at 2.5 mV s^{-1} in 1 mol dm^{-3} KOH. A thin surface film (A) is observed above the InP substrate (B). The surface of the sample has been capped by the amorphous deposit marked at C.

The mechanism by which a porous region can form within the substrate by electrochemical oxidation despite the presence of this dense InP layer at the surface at first sight appears to be inconsistent with the microstructure seen in Fig. 3. The near surface region of the anodized InP, however, was found to contain a low volume fraction of localized channels. Evidence for the existence of these channels was obtained from AFM examination of the surface of electrodes after anodization. Electrodes subjected to potential sweep anodization with upper potentials in the range 0.4 V to 0.53 V were also studied by AFM. Fig. 4 shows AFM images of the surface of an electrode following a potential sweep from 0 V to 0.48 V. The image in Fig. 4(a) clearly shows an etch pit that has formed on the surface. The lower magnification images in Fig 4(b) shows the distribution of these pits on the surface from which estimates were made of the areal density (2.3×10^7 cm^{-2} in this case). It is assumed that both the porous layer and the channels through the near-surface layer are filled with electrolyte, which connects the porous structure with the bulk electrolyte. This enables ionic current to flow and electrochemical oxidation of InP to proceed, thus providing a mechanism by which the porous layer can grow. The channels contained within the near-surface layer can thus be seen to play a critical role in the formation of the nanoporous structure.

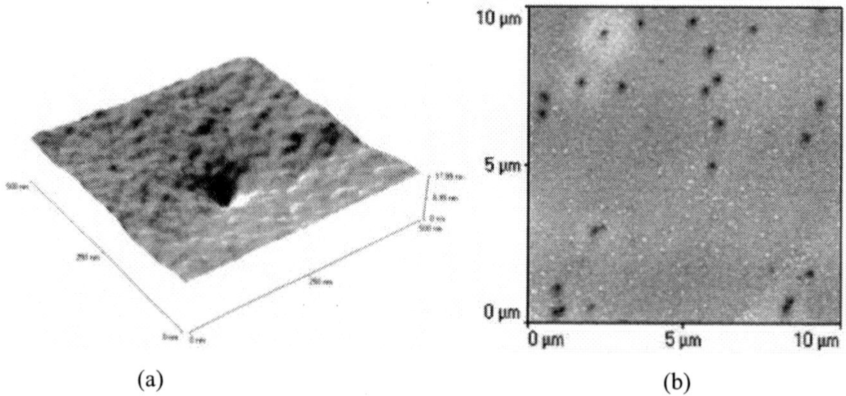

(a)   (b)

Fig. 4 AFM images of the surface of an InP electrode after a linear potential sweep at 2.5 mV s^{-1} in 5 mol dm^{-3} KOH from 0 V to 0.48 V (SCE).

2. Formation of Porous Domains

Fig. 5 shows a typical SEM image of a cleaved β cross section (*i.e.* parallel to the secondary flat) through an electrode after anodization in KOH. The anodization was terminated in the early stages, before a continuous porous layer had developed. Triangular porous regions are clearly visible. This shows that individual, isolated porous domains form in the early stages of anodization. We have shown[2,3] that each such

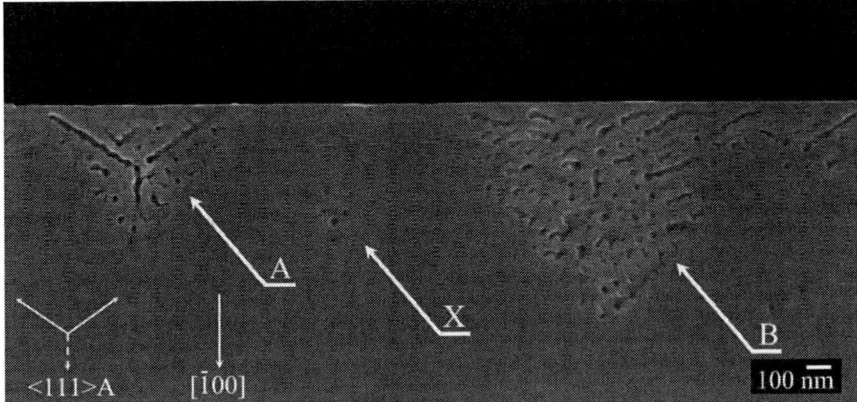

Fig. 5 SEM image of a cleaved β cross section (*i.e.* parallel to the OF flat) of n-InP after an LPS from 0 V to 0.44 V (SCE) in 5 mol dm^{-3} KOH at 2.5 mV s^{-1}. Triangular domain cross sections can be seen at A and B beneath a thin, non-porous, near-surface layer. Both in-plane pores (line-like features approximately along the <111>A directions shown) and through-plane pores (hole-like features) are evident. The small triangular feature at X corresponds to a domain sectioned far from its center as discussed in the text.

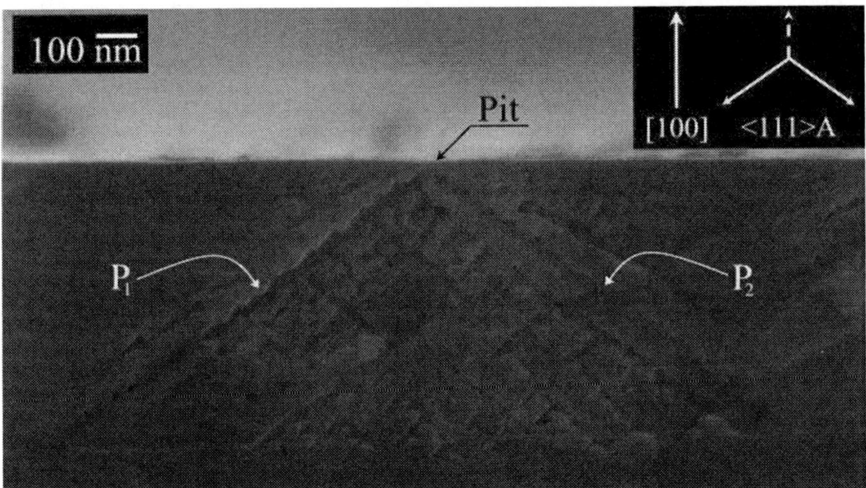

Fig. 6 SEM image of a cleaved α cross section (*i.e.* parallel to the IF flat) of n-InP after anodization under the same conditions as in Fig. 5. A trapezoidal domain cross section can be seen beneath a thin, non-porous, near-surface layer. Both in-plane pores (line-like features approximately along the <111>A directions shown) and through-plane pores (hole-like features) are evident.

domain is separated from the surface by a thin non-porous layer and is connected to the electrolyte by a pit which penetrates this near-surface layer. Clearly, the β-plane cross sections through these domains are triangular in shape. Most of the pores in Fig. 5 apparently pass through the plane, appearing as hole-like features, but several pores appear to be in the plane.

Fig. 6 shows a typical SEM image of a cleaved α cross section through the same electrode as in Fig 5. A quadrilateral porous region with the shape of a truncated isosceles triangle is observed. Two sides of the quadrilateral are parallel to the surface of the wafer, with the shorter side close to the surface (top). We will refer to this shape as an isosceles trapezoid. Many pores appear along lines in the plane but some pores passing through the plane appear as holes. The center of the trapezoid's short side is connected by a short channel through the near-surface layer to a pit in the surface. From this channel an in-plane pore P_1 extends to the bottom left-hand corner of the trapezoid and another P_2 extends symmetrically to the right. Each of these pores makes an angle of $\sim 35°$ with the surface and so is along a <111> direction.

As indicated in the experimental section, the {111} plane between the primary flat (α cleavage plane) and the surface (γ plane) of the wafer is a {111}A plane. Likewise the {111} plane between the secondary flat (β cleavage plane) and the surface is a {111}B plane. The corresponding directions (*i.e.* the normal vectors to these planes) are shown schematically in Fig 1. It is clear from Fig. 1 that the <111>A vectors point downwards from the wafer surface in the α planes and upwards towards the wafer surface in the β planes. Thus P_1 and P_2, which originate at the surface and propagate downwards in an α plane, propagate along <111>A directions (as shown in the inset of Fig. 6). Most of the pores in the region below P_1 and P_2 in the image are parallel to P_1 or P_2 while most of the pores above P_1 and P_2 pass through the plane (*i.e.* appear as holes). Revisiting Fig. 5, the pores in the plane of the image also appear to be along <111> directions. These correspond to holes in Fig. 6. Conversely, in-plane pores in Fig. 6 correspond to holes in Fig. 5.

3. Pore Patterns and Shape of Porous Domains

As already discussed, Figs. 5 and 6 suggest that pores are oriented along <111> directions. If pores originating at a surface pit propagated along all eight <111> directions, the domain formed would have the shape of half a cube. However, as already noted, Figs. 5 and 6 show that α and β cross sections are not rectangular. Furthermore, Fig. 6 shows that primary pores propagate along <111>A directions. We therefore suggest that all pores propagate only along the four <111>A directions. This is expected since the fastest etching planes in InP are generally found to be {111}B, *i.e.* the direction of propagation of the etch front is generally <111>A.

The four <111>A directions may be represented by a set of tetrahedrally symmetrical vectors. To model the domain shape that would result from <111>A pore propagation, we first consider the hypothetical case of pores originating from a point in the bulk of the crystal and propagating along four tetrahedrally symmetrical directions (the vectors *a*, *b*, *c*, and *d*) at rates that are equal at any instant in time. After time t, these four pores will have reached points which form the vertices of a tetrahedron. Branching along the <111>A directions from the primary pores will lead to secondary pores and additional branching will lead to tertiary and quaternary pores that fill the volume of the

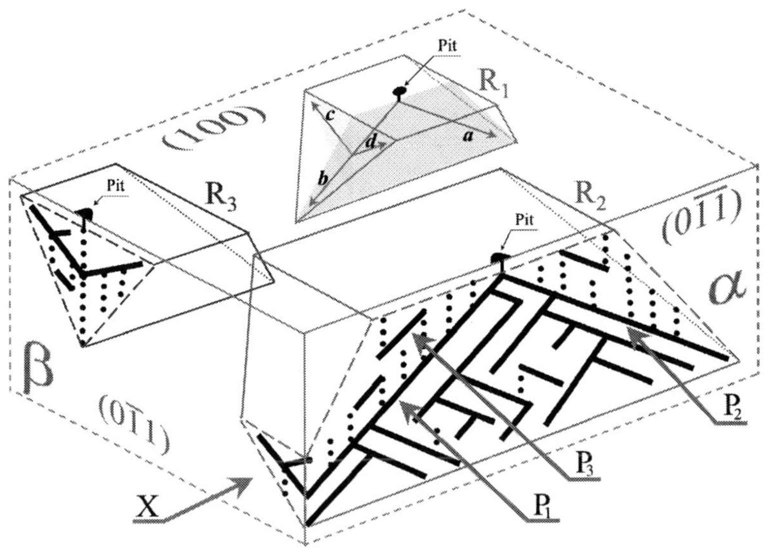

Fig. 7. Schematic of porous domains R_1, R_2 and R_3 beneath a (100) surface and cross sections in the α and β planes. Domains are separated from the surface by a thin layer and each domain originates from a pit penetrating this layer. Pores are represented in R_1 propagating along <111>A directions: downwards along a and b in the α plane and upward along c and d in the β plane. The α cross section intersects R_2 through its center; the β cross section intersects R_3 through its center and R_2 near its vertex (at X). In the cross sections, in-plane pores are shown as solid lines and through-plane pores as small filled circles. P_1, P_2, and P_3 indicate in-plane α pores.

tetrahedron. Thus, such pore propagation along <111>A directions from a point in the bulk of the crystal would lead to a tetrahedral porous domain.

Of course, pores originate near the surface rather than in the bulk of the crystal and therefore the porous domain formed will have the shape of a tetrahedron truncated by a plane parallel to the surface of the crystal at which the pores originate. The schematic in Fig. 7 shows three such porous domains R_1, R_2 and R_3 beneath a (100) surface (γ plane). It also shows the pit from which the domain has formed in each case and the cross sections in α and β planes. As discussed above, α pores propagate downwards along two of the four <111>A directions a and b, and β pores branching from these propagate upwards along the other two <111>A directions c and d. This is shown schematically in R_1. The α cross section shown intersects the domain R_2 through its center forming a trapezoidal shape. Pores originating at the surface pit are shown as solid lines extending downwards in diverging directions to the vertices of the domain. Secondary pores in the plane branch from these primary pores; all pores in the plane are parallel to either a or b, *i.e.* they are α pores. This is consistent with the SEM image in Fig. 6 where two such primary pores, P_1 and P_2, extend from a surface pit to the lower vertices of the domain and many other pores are observed in the plane branching in directions parallel to P_1 or P_2.

We note that no point in the region between the primary pores (P_1 and P_2) and the surface is accessible by paths involving only α pores because these propagate downwards only. However, β pores can branch from them and propagate upwards towards the surface in the orthogonal (*i.e.* β) planes. Subsequent branching along β directions allows these pores to thread back through the region between the primary pores and the surface. In fact, this is the only way that pores can reach that region. The β pores crossing the α cross section are shown as small filled circles in Fig. 7. Such pores can then branch again to give rise to further α and β pores. This explains the observation that pores above P_1 and P_2 in Fig. 6 mainly pass through the plane (β pores) while pores below P_1 and P_2 are mainly in the plane (α pores). Clearly, the detailed features in Fig. 6 resemble those in the schematic α cross section in Fig. 9 and so are consistent with the model.

The β cross section shown in Fig. 7 intersects the domain R_3 through its center forming a triangular shape as discussed above. Pores in the plane (β pores) are shown as solid lines extending upwards from where the α pores from which they branched cross the plane. Pores propagating upwards in a β plane can branch downwards in several α planes to form in each case a pattern of pores that is generally similar to that in the α cross section shown. These pores appear as holes where they pass through the β cross section and are shown as small filled circles. The detailed features in Fig. 5 resemble the schematic β cross section: pores appear mainly as hole-like images with fewer in-plane pores. The β cross section shown also passes through R_2 but does not intersect its base.

4. A Three-Step Model for Pore Formation

The formation of crystallographically oriented (CO) <111>A pores suggests that etching is controlled by the relative rates of the surface reactions at different facets: any satisfactory model of electrochemical pore formation must explain how this can occur even though the rate-determining process (hole generation) occurs only at pore tips. To reconcile these requirements, we propose a three-step model. The three steps (which are illustrated schematically in Fig. 8) are: (1) hole generation at pore tips, (2) hole diffusion and (3) electrochemical oxidation of the semiconductor to form etch products.

Step 1 is the generation of holes at pore tips. As discussed above, this is rate-determining and occurs under the influence of a high electric field due to the small radius of curvature at the pore tip. If each hole formed were to be immediately consumed in an electrochemical reaction, then the resulting etching would be confined to a very small area at the pore tip. This would tend to further sharpen the pore tip and therefore further hole generation and etching reaction would also be confined to that specific site. The

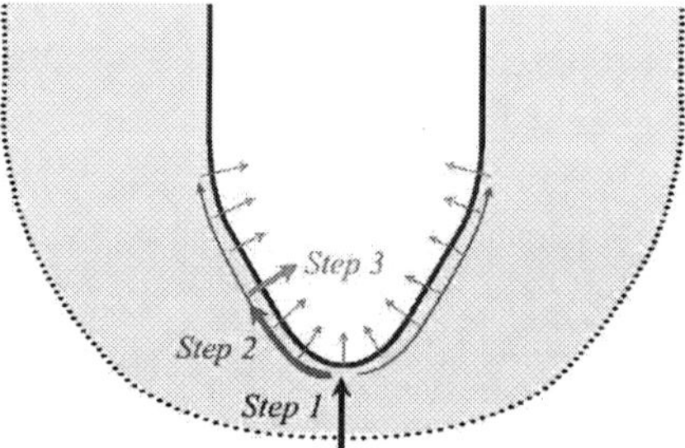

Fig. 8. Schematic representation (not to scale) of a pore near its tip showing the three steps of the model of competitive kinetics: (Step 1) hole generation at pore tips by tunneling of carriers (holes) across the depletion layer (shaded region); (Step 2) hole diffusion at the surface; and (Step 3) electrochemical oxidation of the semiconductor to form etch products.

question therefore arises as to how pore propagation can occur preferentially along crystallographic directions if the rate-determining step occurs at a non-crystallographically defined site *i.e.* the pore tip.

The answer must lie in Step 2: hole diffusion. If holes diffused only a negligible distance, etching would be confined to the site on the surface where they were created. However, when holes may diffuse parallel to the surface of the semiconductor, the electrochemical etching reaction may occur some distance from the pore tip where the holes are created and, as discussed below, this may lead to CO pore formation. Such diffusion of holes from their points of generation to the points where etching ultimately occurs has been reported[25] for photoanodic etching of n-type III-V semiconductors where etching was observed to extend beyond the illuminated region.

Step 3 is the actual electrochemical reaction itself. While the detailed chemistry and mechanism of this have not yet been elucidated, it involves oxidation of InP to indium and phosphorus species dissolved in the electrolyte within a pore. The kinetics of Step 3 do not determine the overall etch rate of a pore: this is determined by the rate of generation of holes at the pore tip (Step 1). However, competition in kinetics between hole diffusion (Step 2) and electrochemical reaction (Step 3) is the principal factor determining the average diffusion distance of holes.

If the kinetics of Step 3 (oxidation reaction) are slow relative to Step 2 (diffusion), then holes can diffuse a significant distance before being annihilated in the oxidation reaction. Then etching can occur at preferred crystallographic sites, such as phosphorus dangling bonds in InP, within a zone in the vicinity of the pore tip and will lead to pore propagation in preferential directions. On the other hand, if the kinetics of Step 3 were fast relative to Step 2, the diffusion distance of holes would be short as they would be annihilated in the oxidation reaction close to where they were created. In that case, etching would occur close to the site of hole generation rather than at preferred crystallographic sites and so there would then be no preferred crystal direction for pore propagation.

5. Mechanism of Crystallographic Etching of Pores.

When the kinetics of Step 3 are sufficiently slow that holes can diffuse to crystallographically preferred reaction sites, etching will eventually reveal the slowest etching crystal facets – the {111}A facets – to form a pyramidal shape with its apex as the pore tip, as shown in Fig. 9. These {111}A planes are terminated by indium atoms, each bonded to three underlying phosphorous atoms. Removal of an indium atom therefore exposes three phosphorous atoms, each with a dangling bond as shown in Fig. 10a. Such a phosphorus atom is easily etched, breaking a bond to each of three indium atoms, two of which are surface atoms. Each of these now has two dangling bonds (each already had one) and consequently is easily etched, revealing two new phosphorous atoms, each with a dangling bond.

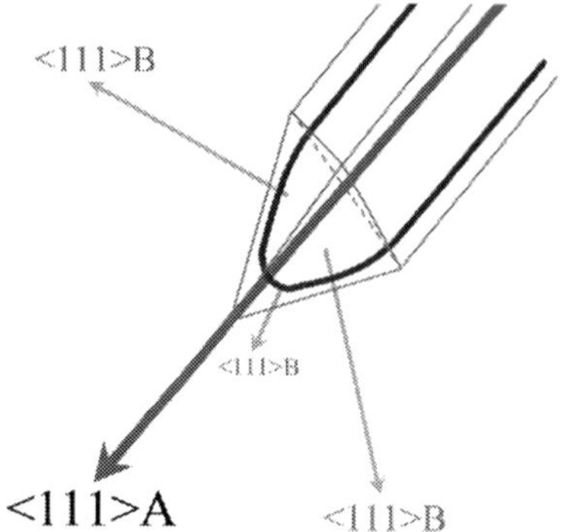

Fig. 9. Schematic representation (not to scale) of a pore formed by the etching mechanism described in Fig. 8. The idealized shape with three {111}A facets is represented by grey lines (———). The actual pore (represented by blue lines (———) will have finite radius of curvature at the tip and will generally have rounded pore walls.

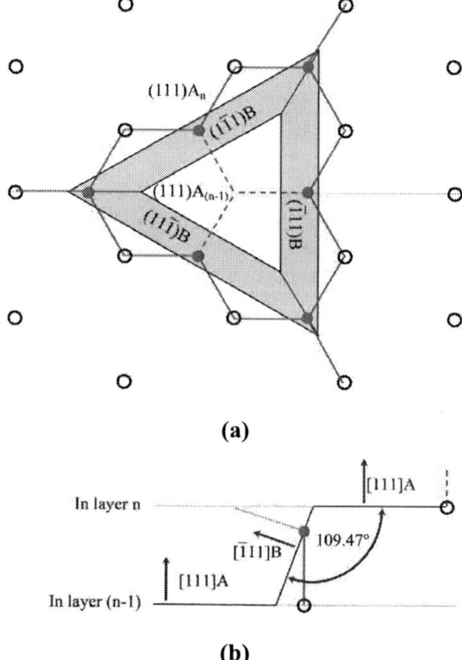

Fig. 10. (a) An In vacancy (·-) on a {111}A surface creating three dangling P bonds (-----). These P atoms (●) correspond to a {111}B monatomic ledge on three sides of the vacancy (projections of which on the {111}A plane are shown). Indium atoms are shown as open circles (O). (b) Cross-sectional view orthogonal to the surface {111}A plane and through the In vacancy and one of the P atoms in a plane containing its dangling bond. The surface {111}A plane (n) and the next underlying {111}A plane (n-1) are shown. The {111}B ledge is represented at an angle of 109.47° to the surface plane and contains the P atom with its dangling bond towards the In vacancy and normal to the surface of the ledge. Note that the P atom is bonded to an In atom in the n-1 plane; as can be seen in (a) it is also bonded to two In atoms in the n plane.

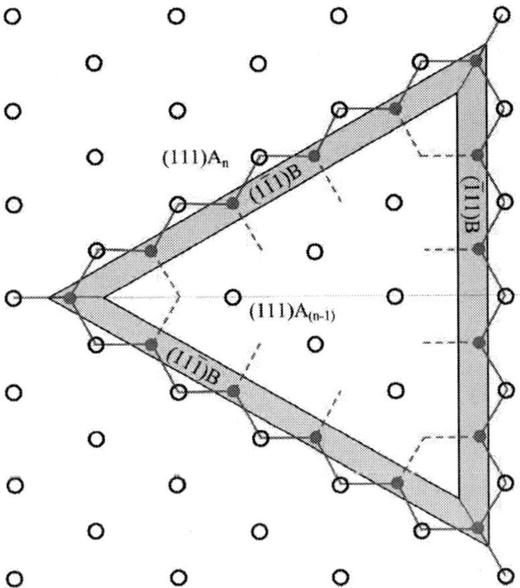

Fig. 11. Expanded three-sided region as etching progresses further away from the original In vacancy shown in Fig. 10. This occurs as follows: as the P atoms on the ledges in Fig. 5a are etched, In atoms with two dangling bonds are exposed which in turn etch. The corner P atoms (which originally had no dangling bonds) are then exposed and etch. Thus each ledge advances to the next row of P atoms.

Phosphorous atoms with single dangling bonds can be considered to be part of a {111}B surface oriented normal to the bonds.[26] Thus, the removal of a single indium atom from a {111}A surface creates three monatomic ledges with {111}B faces, each at an angle of 109.47° to the surface as shown in Fig. 10b. Etching of the phosphorous atoms on these ledges, and the associated indium atoms, causes the ledges to advance, consequently increasing the size of the three sided region where the next underlying {111}A plane is exposed (Fig. 11).

New vacancy sites are not formed easily but, once formed, they expand as in Fig 11 by rapid two-dimensional etching along the surface to expose the next {111}A plane. Eventually, an indium vacancy forms in the newly exposed {111}A face and the etching process continues to expose the next underlying {111}A plane (see Fig. 10). Thus, in the vicinity of the pore tip, the {111}A faces are etched, one monolayer at a time, so that the tip region maintains its pyramidal shape, and so remains sharp, as it propagates. Symmetrical etching of the three {111}A faces forming the tip causes it to propagate in the (fourth) <111>A direction.

The crystallographic etching that maintains {111}A facets in the vicinity of the tip is controlled by the supply of holes diffusing from the tip. The hole concentration at the surface of the facets decreases with distance from the tip, both because of divergence of the diffusing flux and because of consumption of holes in the oxidation reaction. Thus, at some distance from the tip the rate of etching becomes negligibly small. At intermediate distances, a transition region from etching to non-etching must therefore exist where propagation of ledges on individual planes ceases and where the surface curves away from the {111}A facets towards the central axis of the pore.

Thus, as a pore etches multiple ledges are propagating outwards from multiple initiation points (indium vacancies) on each of the facets. Where these propagating ledges meet they can form three-sided corners at which the local electric field is sufficiently high that hole generation can occur. These sites can become new pore tips when they develop in locations sufficiently far from the depletion regions around adjacent pores. In this way, branching of pores can occur along any of the <111>A directions.

In summary, the three-step model can explain how the concepts of CO etching and rate control by localized hole generation at pore tips can be reconciled. If Step 3 (oxidation reaction) is sufficiently fast that Step 1 (hole generation) is rate-determining but sufficiently slow in comparison with Step 2 (diffusion) that holes can diffuse a significant distance, then preferential etching of {111}B faces can occur. This leads to the formation of a faceted pore tip which, through etching of {111}A planes one monolayer at a time, will lead to the propagation and branching of pores along <111>A directions.

CONCLUSIONS

Anodization of highly doped (10^{18} cm^{-3}) n-InP in $2 - 5$ mol dm^{-3} KOH under potentiostatic or potentiodynamic conditions results in the formation of a nanoporous sub-surface region beneath a thin (typically ~40 nm) dense near-surface layer. LSVs show a pronounced anodic peak, typically at 0.48 V for a 2.5 mV s^{-1} scan in 5 mol dm^{-3} KOH, corresponding to the formation of the porous region. However, no porous regions were formed during anodization in 1 mol dm^{-3} KOH. For upper potentials in the range 0.4 V to 0.53 V, AFM images clearly show etch pit formation on the surface.

Detailed TEM and SEM studies show that pores originate from surface pits and an individual, isolated porous domain is formed beneath each pit in the early stages of anodization. Each such domain is separated from the surface by a thin non-porous layer and is connected to the electrolyte by its pit, which penetrates this near-surface layer. Pores emanate from these points along the <111>A crystallographic directions to form domains with the shape of a tetrahedron truncated symmetrically through its center by a plane parallel to the surface of the electrode. Cross sections of these domains are trapezoidal and triangular, respectively, in the α and β cleavage planes of the wafer. The observed SEM and TEM cross sections show pore patterns that are in good agreement with those predicted.

We propose a three-step model of electrochemical pore formation that explains how crystallographically oriented etching can occur even though the rate-determining process (hole generation) occurs only at pore tips. Step 1 is the generation of holes at pore tips under the influence of a high electric field due to the small radius of curvature at the pore tip. Step 2 is the diffusion of holes parallel to the surface of the semiconductor, enabling the electrochemical etching reaction to occur some distance from the pore tip where the holes are created. Step 3 is the actual electrochemical reaction itself.

Step 1 determines the overall etch rate. However, competition in kinetics between hole diffusion (Step 2) and electrochemical reaction (Step 3) determines the average diffusion distance of holes and this in turn determines whether etching is crystallographic. If the kinetics of Step 3 are slow relative to Step 2, then etching can occur at preferred crystallographic sites, such as phosphorus dangling bonds in InP, within a zone in the vicinity of the pore tip and this will lead to pore propagation in preferential directions.

Under these conditions, etching will eventually reveal the slow etching {111}A facets to form a pyramidal pore tip. Indium vacancy sites formed on these facets expand by rapid two-dimensional etching along the surface to expose the next {111}A plane. Thus, in the vicinity of the pore tip, the {111}A faces are etched, one monolayer at a time, so that the tip region maintains its pyramidal shape, and so remains sharp, as it propagates. Symmetrical etching of the three {111}A faces forming the tip causes it to propagate in the (fourth) <111>A direction. As a pore etches, propagating atomic ledges can meet to form sites that can become new pore tips and this enables branching of pores along any of the <111>A directions.

Acknowledgements

R. P. Lynch and N. Quill would like to thank the Irish Research Council (IRC) for PhD scholarships to perform this research. R. P. Lynch acknowledges a joint IRC - Marie Curie Fellowship under grant no. INSPIRE PCOFUND-GA-2008-229520. The authors would also like to acknowledge the support of the Tyndall National Institute. This support was provided through the SFI-funded National Access Programme (Project NAP No. 37 and 70). The authors also acknowledge many contributions to the electron microscopy by S. B. Newcomb and to the AFM measurements by M. Serantoni.

REFERENCES

1. R.P. Lynch, N. Quill, C. O'Dwyer, S. Nakahara and D.N. Buckley, *Phys. Chem. Chem. Phys.*, **15**, 15135-15145 (2013)
2. R.P. Lynch, C. O'Dwyer, N. Quill, S. Nakahara, S.B. Newcomb, and D.N. Buckley, *J. Electrochem. Soc.*, 160, D260-D270 (2013).
3. C. O'Dwyer, D.N. Buckley, D. Sutton, M. Serantoni, and S.B. Newcomb, *J. Electrochem. Soc.*, **154**, H78 (2007)
4. C. Fang, H. Föll, J. Carstensen and S. Langa, *phys. stat. sol. (a)* **204** (5), 1292 (2007)
5. C. O'Dwyer, D.N. Buckley, D. Sutton, and S.B. Newcomb, *J. Electrochem. Soc.*, **153**, G1039 (2006)
6. L. Santinacci, I. Gerard, M. Bouttemy, M. Marques and A. Etcheberry, *ECS Trans.*, **16**, 411 (2008).
7. J.J. Kelly and H.G.G. Philipsen, *Curr. Opin. Solid-State Mater. Sci.* **9**, 84 (2005)
8. S. Langa, J. Carstensen, I.M. Tiginyanu, M. Christopersen and H. Föll, *Electrochem. Solid-State Lett.* **4**, G50 (2001)
9. A. M. Goncalves, L. Santinacci, A. Eb, I. Gerard, C. Mathieu and A. Etcheberry, *Electrochem. Solid-State Lett.*, **10**, D35 (2007).
10. D.J. Lockwood, P. Schmuki, H.J. Labbé, J.W. Fraser, *Physics E*, **4**, 102 (1999)
11. T. Osaka, K. Ogasawara and S. Nakahara, *J. Electrochem. Soc.*, **144**, 3226 (1997)
12. A.G. Cullis, L.T. Canham and P.D.J. Calcott, *J. Appl. Phys.*, **82**, 909 (1997)
13. M. Schoisswohl, J.L. Cantin, M. Chamarro, H. J. von Bardelben, T. Morganstern, E. Bugiel, W. Kissinger and R.C. Andreu, *Phys. Rev. B*, **52**, 11898 (1995)
14. B.H. Erné, D. Vanmaekelbergh, J. J. Kelly, Adv. Mater., 7, 739 (1995)

15. T. Takizawa, M. Nakahara, E. Kikuno and S. Arai, J. Electron. Mat., 25, 657 (1996)

16. S. Langa, J. Carstensen, M. Christophersen, K. Steen, S. Frey, I.M. Tiginyanu and H. Föll, *J. Electrochem. Soc.* **152**, C525 (2005)

17. N. Quill, R. P. Lynch, C. O'Dwyer, D. N. Buckley, *ECS Trans.*, **50(37)**, 143 - 153 (2013) *doi:10.1149/05037.0143ecst*

18. H. Fujikura, A. Liu, A. Hamamatsu, T. Sato and H. Hasegawa, Jpn. J. Appl. Phys. 39, 4616 (2000)

19. S. Langa, I.M. Tiginyanu, J. Carstensen, M. Christopersen and H. Föll, *Electrochem. Solid-State Lett,* **3**, 514 (2000)

20. F.M. Ross, G. Oskam, P.C. Searson, J.M. Macaulay and J.A. Liddle, *Philos. Mag. A* **75**, 525 (1997)

21. H. Tsuchiya, M. Hueppe, T. Djenizian and P. Schmuki, *Surf. Sci.* **547,** 268 (2003)

22. X. G. Zhang, *J. Electrochem. Soc.,* **151**, C69 (2004)

23. S.N.G. Chu, C.M. Jodlauk and W.D. Johnston, Jr., *J. Electrochem. Soc.* **130**, 2399 (1983)

24. D. Soltz and L. Cescato, *J. Electrochem. Soc.* **143**, 2815 (1996)

25. F. W. Ostermayer, P. A. Kohl, R. M. Lum, *J. Appl. Phys.,* **58**, 4390 (1985)

26. D. N. MacFayden, *J. Electrochem. Soc.,* **130**, 1934 (1983)

ECS Transactions, 69 (14) 33-48 (2015)
10.1149/06914.0033ecst ©The Electrochemical Society

Towards Electrochemical Fabrication of Free-Standing Indium Phosphide Nanofilms

Nathan Quill,[a] Colm O'Dwyer,[b] D. Noel Buckley [a] and Robert P. Lynch [a,†]

[a] Department of Physics & Energy, and Materials & Surface Science Institute, University of Limerick, Ireland
[b] Department of Chemistry, and Tyndall National Institute, University College Cork, Ireland

Abstract

The formation of sub-surface truncated tetrahedral voids beneath suspended ~40 nm thick dense InP shelves is achievable via a two-step etching method. The first step involves the electrochemical anodisation of n-type InP in aqueous KOH electrolyte resulting in the formation of truncated tetrahedral domains of pores beneath an ~40 nm thick dense surface layer with an individual pit that penetrates the dense surface layer of each domain. The second step involves the preferential chemical etching of these porous domains with only limited etching of the surrounding bulk InP and dense surface layer. The resulting structure of each domain is a truncated tetrahedral void with a dense InP layer suspended above it. This new technique may be a very useful tool in the fabrication of devices based on III-V semiconductors and it may be possible to extend the technique to the fabrication of free-standing InP nanofilms.

Introduction

III-V semiconductors have received considerable research attention due to their use in the optoelectronics industry. In recent years, particular interest has been paid to the fabrication and characterisation of nano-scale structures for various applications. Nanopore formation due to electrochemical etching has been demonstrated in a range of semiconductors from Si to InP(1-12). The morphology of these porous layers can vary wildly between different semiconductors(13), and for an individual semiconductor with the variation of temperature(14), composition(15, 16) and concentration(17) of electrolyte, and orientation(18) and doping density(19) of the substrate. Many different pore morphologies have been observed and several models have been proposed(20-32) to explain them. However, it is generally accepted that the propagation of nanopores in highly doped n-type semiconductors is controlled by hole generation under the influence of a high electric field due to the small radius of curvature at the pore tip(29, 30).

We have previously investigated(14,15,33-39) the early stages of pore formation in InP in aqueous KOH electrolytes. Pores originate from etch pits in the electrode surface and propagate along <111>A directions to form porous domains beneath a thin dense (*i.e.* low

† E-mail: Robert.Lynch@UL.ie

porosity) surface layer penetrated only by the surface pits(35). The domains shape is that of a truncated tetrahedron due to the preferential growth direction of the individual pores. Eventually these domains merge to create a continuous porous layer(39).

There have been a number of applications proposed for the various porous structures that can be formed(40-53) in different semiconductors under different conditions. In this paper we demonstrate how the porous InP domains that we have previously characterised can be chemically etched, leaving behind a unique structure: a truncated tetrahedron shaped void, below the dense surface layer which is ~40 nm thick and freely suspended above the void.

Experimental

Wafers were monocrystalline, sulfur-doped, n-type indium phosphide (n-InP) grown by the liquid-encapsulated Czochralski (LEC) method and supplied by Sumitomo Electric. They were polished on one side and had a surface orientation of (100) and a carrier concentration in the range $3-7 \times 10^{18}$ cm^{-3}. To fabricate working electrodes, wafers were cleaved into coupons along the natural {011} cleavage planes. Ohmic contact was made by alloying indium to the back of a coupon. The electrodes were cleaned by rinsing and then immersion in firstly acetone, then methanol and finally water. The back and the cleaved edges were then isolated from the electrolyte by means of a suitable varnish. The electrode area was typically 0.2 cm^2. Prior to immersion in the electrolyte, the working electrode was immersed in a piranha etchant (3:1:1 H_2SO_4:H_2O_2:H_2O) for 4 minutes and then rinsed with deionized water.

As described previously(14,15,33-39) anodisation was carried out in aqueous 5 mol dm^{-3} KOH electrolytes in the absence of light at room temperature ~19°C using a linear potential sweep (LPS). A conventional three-electrode cell configuration was used, employing a platinum counter electrode and a saturated calomel electrode (SCE) to which all potentials are referenced. LPSs were carried out at 2.5 mV s^{-1} from 0.0 V to an E_{max}. A CH Instruments Model 650A Electrochemical Workstation interfaced to a Personal Computer (PC) was employed for cell parameter control and for data acquisition.

Cleaved ($0\bar{1}1$) and ($0\bar{1}\bar{1}$) cross sections of electrodes were examined using a Hitachi S-4800 field-emission scanning electron microscope (FE SEM) operating at 5 kV, unless otherwise stated.

Results and Discussion

Formation of Unique Porous Structure

We have previously reported the formation of porous layers in n-InP that has been anodised in > 2 mol dm^{-3} KOH solutions(14,15,33-39). A typical linear sweep voltammogram (LSV) obtained during the potentiodynamic formation of porous InP in 5 mol dm^{-3} KOH is shown in Fig. 1. On the plot, three potentials are labelled: E_{pit}, the pitting potential, E_{p1}, the potential of the first current peak, and E_{p2}, the potential of the second current peak. Since a depletion layer exists at the n-InP surface under the conditions required for pore formation, the etch rate is generally limited by hole supply at the semiconductor surface. In the absence of illumination or a highly oxidising species within the electrolyte, these holes must be generated within the bulk semiconductor. In order for these holes to reach the surface of the semiconductor and facilitate etching, they must cross the depletion layer at the surface. The electric field enhancement that occurs at a sharp etch pit or pore tip(29, 30) can be sufficient to allow a hole to tunnel across the depletion layer once the potential drop across the depletion layer is sufficiently large. For this reason, at potentials below E_{pit}, no significant current flows and no significant electrode modification takes place.

As the current begins to increase at potentials greater than E_{pit}, etch pits begin to appear in the electrode surface. Figure 2 shows an SEM image of the (100) surface (the surface exposed to the electrolyte) of a porous InP sample. A number of these etch pits can be seen penetrating the InP surface. It is noted however, that the actual surface remains mostly intact *i.e.* the surface porosity is quite low.

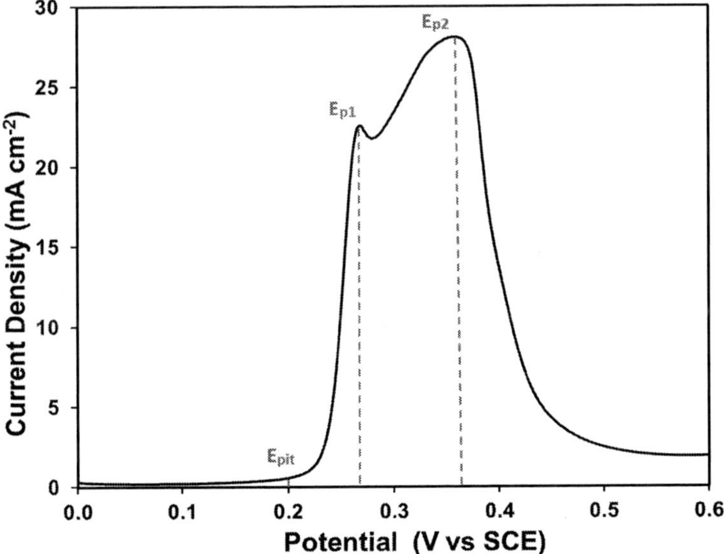

Fig. 1 LSV of InP in 5 mol dm^{-3} KOH. The potential was swept from 0.0 to $E_{max} = 0.6$ V (SCE) at 2.5 mV s^{-1}. Marked on the LSV are E_{pit}, the potential at which the first etch pits start to appear on the electrode surface, E_{p1}, the potential of the first current peak, and E_{p2}, the potential of the second current peak.

Due to limitations in carrier supply(31), etching is initially constrained to be normal to the electrode surface as was seen for the surface pits in Fig. 2. However, once the etch front has progressed by approximately the thickness of the depletion layer, pores begin to branch out from this initial pit and propagate along <111>A crystallographic directions(31, 33).

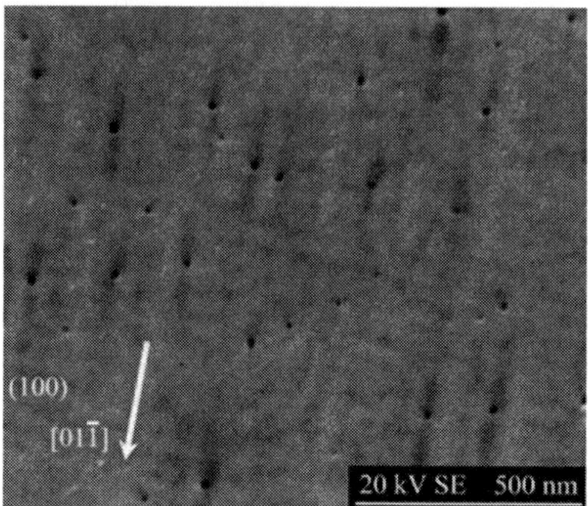

Fig. 2 SEM micrograph of an InP (100) surface following an LPS from 0.0 to 0.537 V (SCE). A number of etch pits are clearly visible on the electrode surface. Since the image was taken at 20 kV both the surface pits and some sub-surface features are shown.

Once crystallographically oriented pores have formed from the original etch pit, these pores continue to grow and branch, forming a complex network filled with electrolyte and etch products. As the pores branch, the number of actively growing pore tips increases, leading to a rapid increase in current in the LSV after E_{pit}. Each pore network may contain many actively growing (etching) pore tips but they are all connected to the bulk electrolyte via the original surface pit from which the initial pores formed. We call such a pore network (with only a single connection to the bulk electrolyte) a *porous domain*. These porous domains have the characteristic shape of a truncated tetrahedron(33) (due to their crystallographically oriented etching). This is illustrated schematically in Fig. 3 which shows the overall shape of a porous domain formed by pores etching exclusively along <111>A directions. This schematic also shows the porous domain cross sections that would be expected on different crystallographic planes. On the ($0\bar{1}1$) plane an inverted triangle or "v-groove" shaped cross-section is expected. On the ($0\bar{1}\bar{1}$) plane a trapezoidal or "dove-tail" cross-section is expected.

The equivalent SEM cross-sections of actual porous domains are shown in Fig. 4. A ($0\bar{1}1$) cross section is shown in Fig. 4a. It shows the characteristic v-groove domain shape. A similar domain, viewed on the ($0\bar{1}\bar{1}$) plane is shown in Fig. 4b. It shows the characteristic dove-tail domain shape. Both of these images also show that the porous domain is separated from the semiconductor surface by a surface layer of dense (*i.e.* low porosity) InP (labelled as A in each micrograph) which is penetrated at only one point, by the surface pit.

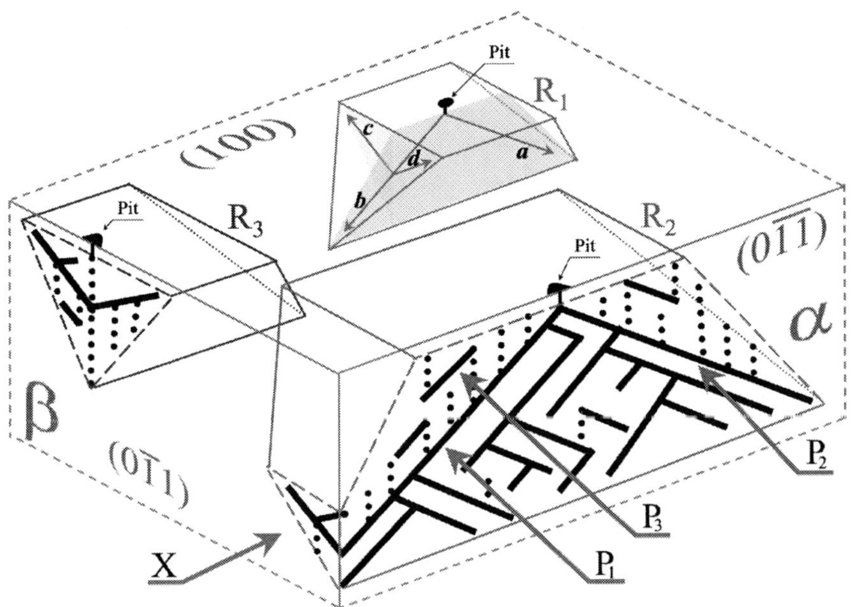

Fig. 3 Schematic of porous domains R_1, R_2 and R_3 beneath a (100) surface and cross sections in $(0\bar{1}1)$ and $(0\bar{1}\bar{1})$ planes. Domains are separated from the surface by a thin layer of dense InP and each domain originates from a pit penetrating this layer. Pores are shown oriented along <111>A directions and the unique porous domain shapes that are seen in SEM cross sections are also shown.

As etching continues, these domains eventually merge with each other, which slows the increase in current, eventually resulting in a current peak at E_{p1}. Where the domains merge, the depletion layers surrounding individual pores overlap, resulting in a cessation of etching along that direction. Once all of the individual domains merge, a continuous porous layer is formed, which coincides with the trough in current seen just after E_{p1} in Fig. 1. However, pores at the base of the porous layer can continue to grow deeper in to the substrate, which results in the gradual thickening of the porous layer. This results in the pseudo-linear increase in current seen between the two current peaks in Fig. 1. This thickening continues until a second current peak is observed at E_{p2}, which is quickly followed by the cessation of porous layer growth(39).

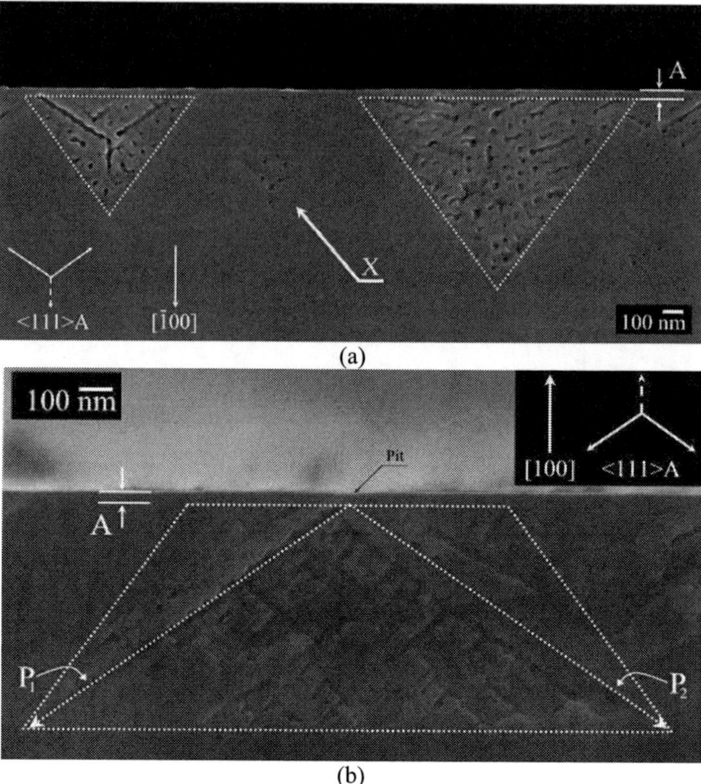

Fig. 4 SEM images of n-InP after an LPS from 0 V to 0.44 V (SCE) in 5 mol dm⁻³ KOH at 2.5 mV s⁻¹. (a) A cleaved (0Ī1) plane in which a characteristic v-groove porous domain cross section can be seen beneath a thin dense surface layer at A. (b) A cleaved (0ĪĪ) plane showing a characteristic dove tail domain cross section beneath a thin dense surface layer at A.

We have previously described how etching of porous layers in InP in KOH is a purely electrochemical process and this leads to some features which are not present in many other pore systems(31). As mentioned earlier, holes are supplied from the bulk of the semiconductor leading to narrow etch pits which penetrate a layer of dense InP. Since no chemical etching takes place, this surface layer of dense InP remains intact throughout the etching process. This was shown in Fig. 2 where the surface of a porous InP layer was shown to be almost completely intact, with the occasional surface pit offering the only hint at the sub-surface porosity. This creates a unique structure which consists of a porous layer sandwiched between two layers of dense InP; the dense surface layer and the bulk semiconductor. Furthermore, the structure obtained can be tailored by controlling the upper potential E_{max} in the LPS. In this way, isolated porous domains can be formed if

E_{max} is chosen such that $E_{pit}<E_{max}<E_{pl}$. The average size of these domains can also be tailored by stopping the potential sweep at an earlier or later time within this window.

In the next section, all of the porous samples shown have had their potential swept to $E_{pit}<E_{max}<E_{pl}$ as just described, so that the etching of individual porous domains can be studied.

<u>Dissolution of porous layer</u>

It is possible, at an elevated temperature, to chemically etch InP using a mixture of two parts (by volume) H_2SO_4 with three parts water (*i.e.* ~7.5 mol dm^{-3} H_2SO_4). Where the InP being etched has undergone a previous step to form domains of pores beneath the dense surface layer, it is observed that such chemical etching can result in a large number of bubbles emanating from the sample surface. These bubbles do not appear immediately but after a short period of time. SEM investigations of cleaned samples show that these bubbles corresponded to the rapid etching of the skeleton structure of InP in the porous domains. These bubbles could be PH_3 which may signify the progression of enhanced etching of the porous skeleton. Micrographs of the (100) surface of such a sample are shown in Fig. 5. In Fig. 5a an individual domain displaying the characteristic truncated tetrahedral domain shape can be seen. However in Fig. 5b two intersecting domains can be seen. In both images not alone has the internal InP structure been etched but the surface layer of dense InP has also been etched. This etching has made the truncated tetrahedral structures of the domains visible. Therefore, in Fig. 5a the elongation of the domain along one of the two <011> axes and the pointed vertices that are seen as symmetric sub-surface features are displayed. Where truncated tetrahedral shapes of domains are merged together along the [01$\bar{1}$] axis there is a thick region (i.e. of up to several hundred nanometres, as shown at W in Fig. 5c) of dense InP that is between their point of merging and beneath the electrode-surface. Therefore when the porous skeleton and the dense surface layer have been etched, this thick region of dense InP becomes suspended above a void, as can be seen in Fig. 5b.

It can be seen in Fig. 5a and Fig. 5b that in both of the images the openings to the surface are not square, like the shape of the truncated domains just beneath the surface, but almost hexagonal. This may be due to etching outwards from the internal surfaces of the pores in the domains. It follows, that since the two primary pores grow downwards along the <111>A directions (towards the vertices of the domain) the surface pit should widen to a greater extent along these directions where there is enhanced etching due to the chemistry within the pores. Hence, a greater degree of etching results along the [01$\bar{1}$] axis and void openings in the surface are hexagonal shaped.

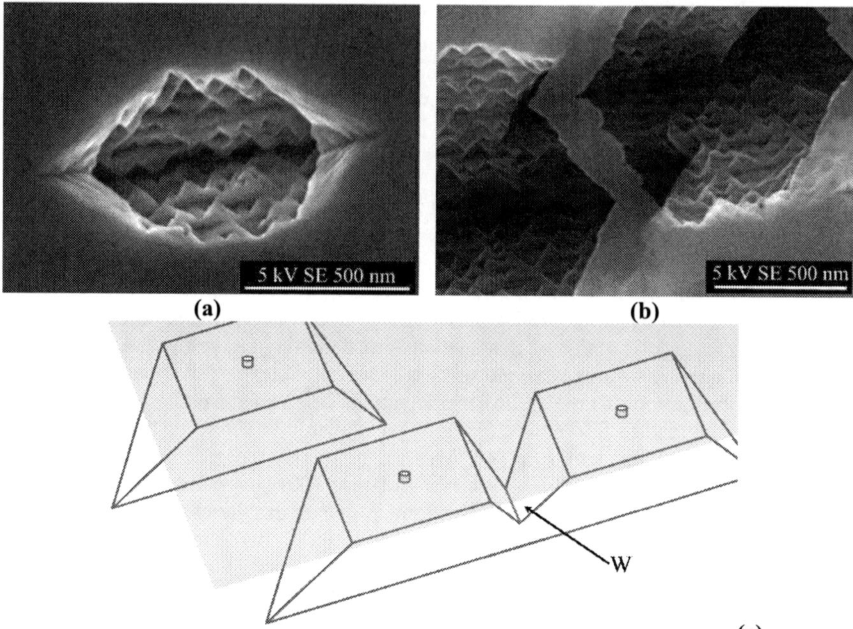

Fig. 5 SEM micrographs of (a) an individual porous domain and (b) two intersecting domains that have been etched by ~7.5 mol dm^{-3} H$_2$SO$_4$ at 70°C for 150 s. The sample was taken directly from room temperature water and placed in the H$_2$SO$_4$ solution. (c) A schematic representation of the individual and intersecting porous domain outlines prior to etching is shown and a volume of bulk InP that is surrounded due to the merging of two domains is indicated at W.

Where the InP being etched contains porous domains the inter-pore regions (pore walls) etch much more quickly than the bulk. The enhanced rate of etching within the porous structure may be due to the combination of two effects. The first effect is due to the surface-area to volume ratio being greater for the pore walls than for the dense surface layer of InP. This results in the build-up of a large number of chemical products within the porous domain. The second effect may therefore be due to enhancement of the etching rate within the porous domains by the chemical products trapped inside, thus changing the chemistry within the domains. Furthermore, such an effect − that is dependent on an initiation time, during which the reactants required for enhanced etching accumulate − would explain the absence of bubbles when the samples are initially placed in the H$_2$SO$_4$ solution. Therefore it is likely (since only ~20% of the volume of each domain is porous and ~80% of the domain is InP) that the accumulation of products in the enclosed regions of the porous domain results in a change in the chemistry within the domain and a subsequent enhancement in etching of the pore walls.

If it is the case that the porous structures undergo enhanced etching, it follows that control of etching time could lead to structures such as those shown in Fig. 6. In this schematic it is shown that a layer of dense InP, ~40 nm thick, that is suspended over a void can be formed if an etching mechanism preferentially removes the internal domain skeleton while leaving the surface layer intact.

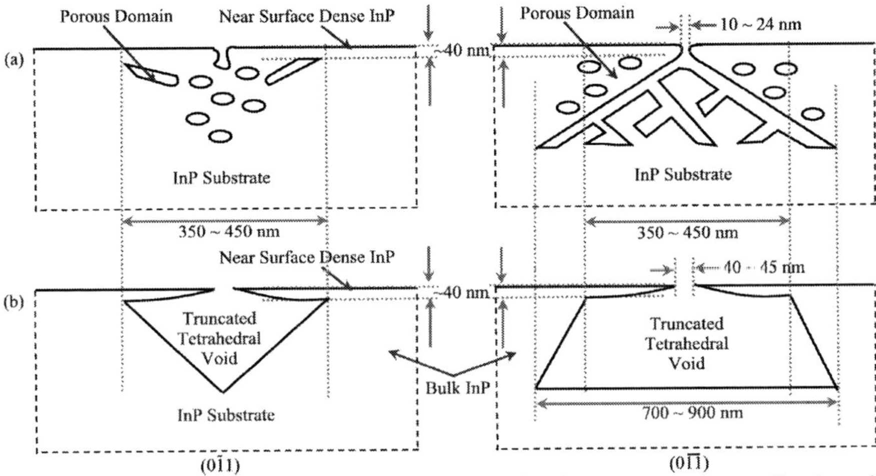

Fig. 6 Schematic showing the layout of (a) a domain of pores grown by anodization of InP and (b) the void created by the chemical removal of the internal InP skeleton of such a domain. Such a process would provide isolated and suspended regions, of thicknesses ~40 nm, above volumes void of semiconductor.

<u>Effect of H_2SO_4 Concentration on Etching</u>

Several concentrations ranging from 1 mol dm^{-3} H_2SO_4 to concentrated H_2SO_4 were tested at room temperature. These experiments were performed so as to tailor the etching of the skeleton of the porous domain with the aim of realizing structures similar to those shown in Fig. 6. During all of these experiments there was no observation of bubbles emanating from the sample surfaces. In the previous experiments, which were carried out at elevated temperatures, the presence of bubbles emanating from the surface coincided with the etching of the skeleton of the porous domain. Therefore the lack of bubbles suggests that no significant etching occurs in H_2SO_4 solutions at room temperature, regardless of concentration. Further examination by SEM surface and cross-sectional analyses confirmed that no noticeable etching of the porous skeleton occurred during the time the samples were held in solution at room temperature. This therefore suggests that the etching mechanism and the production of bubbles are inextricably linked and interdependent. Furthermore, it suggests that elevated temperature is essential for the etching mechanism. A possible reason for this may be that, at low temperatures, the etching of InP by H_2SO_4 solution is slow and the products responsible for the enhanced etching have sufficient time to escape from the confined regions of the porous domains and therefore never accumulate in sufficient concentrations for enhanced etching to occur.

Effect of H_2SO_4 Temperature on Etching

The effect of temperature on the chemical etching of porous structures in concentrated H_2SO_4 solutions (~7.5 mol dm^{-3}) was also investigated. Prior to etching, the samples were made porous through anodizing in 5 mol dm^{-3} KOH via an LPS from 0.0 to E_{max} = 0.25 V (*i.e.* E_{pit}<E_{max}<E_{p1} so that some isolated domains were obtained) at 2.5 mV s^{-1}. These samples were then rinsed with water prior to steeping in water overnight. Each sample was then steeped in H_2SO_4 solution for 90 seconds at bath temperatures ranging from 20°C to 90°C.

Above 60°C bubbles were found to emanate from the sample surface during steeping with a large amount of bubbles emanating above 70°C. As mentioned earlier, the observation of bubbles coincides with the chemical etching of InP. Therefore etching of InP in ~7.5 mol dm^{-3} H_2SO_4 does not seem to occur unless the solution temperature is 60°C or higher. Fig. 7 shows SEM images of a sample etched at 60°C for 90 s. Despite the observation of bubbles during etching, very little etching of the porous domain has occurred. A slight increase in the pore width (from 29 nm to 38 nm) and in the average pit diameter (from 15 nm to 20 nm) is observed in Fig. 7a and Fig. 7b, respectively.

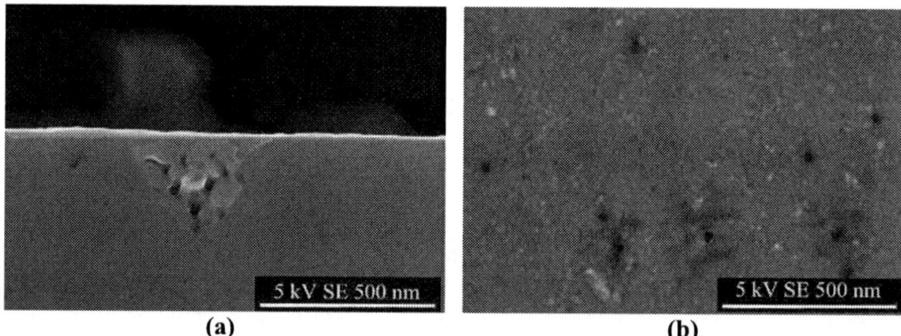

Fig. 7 SEM (a) $(0\bar{1}1)$ cross-sectional and (b) (100) surface views of InP containing porous domains created by LPS in KOH that have been chemically etched in 7.5 mol dm^{-3} H_2SO_4 for 90 s at 60°C.

In contrast to Fig. 7, Fig. 8 shows an SEM image of a sample that has been 'over-etched'. The sample was etched for the same time of 90 s but at a higher temperature of 80°C. As shown in Fig. 8a these conditions result in the removal of most of the dense surface layer leaving only the remnants of the original surface in regions were no porous domains had been formed. This etching of the dense surface layer is also seen in the cross-sectional image of Fig. 8b. It can be observed in this image that the skeletons of the porous domains have been etched away and that the etching has also resulted in the non-uniform etching of the triangular outline of the domains. While an increase in temperature from 60°C to 80°C is expected to lower the energy barrier for chemical reaction and thereby enhance the reaction kinetics, the enormous increase in the amount of etching suggests that the etching process is also being enhanced by the onset of an additional mechanism (as alluded to earlier in the text). If the presence of etch products within the pores is responsible for the enhancement of the reaction, any increase in the etch rate would be self-enhancing, since it would result in the trapping of further chemical products within the porous structure. Therefore it seems very plausible that the rapid increase of etch rate with temperature is due to such a change of chemistry within the domains.

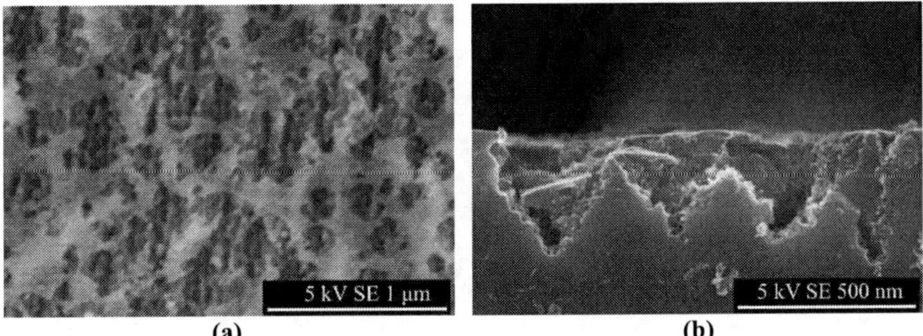

Fig. 8 SEM (a) (100) surface and (b) (0$\bar{1}$1) cross-sectional views of InP containing porous domains created by LPS in KOH that have been chemically etched in 7.5 mol dm^{-3} H$_2$SO$_4$ for 90 s at 80°C.

By choosing an intermediate temperature of 70°C it is possible to etch the internal skeleton structure completely, while leaving the dense surface layer of isolated domains almost fully intact. Fig. 9 is a (0$\bar{1}$1) SEM cross-section from such an experiment. It can be observed that the dense surface layer and the triangular outline of the domain are still intact but that the internal skeleton structure is completely removed. It can also be seen that the etching process is more controlled at this temperature with no over-etching of the domain outline being seen. The majority of the isolated domain cross-sections observed for the experiment shown in Fig. 9 ranged from 350 to 450 nm and measurements of the surface-pit diameter for these domains showed an increase to between 40 and 45 nm (from between 10 and 24 nm prior to etching).This is in contrast with the less controlled etching at the higher temperature (Fig. 8b).

This controlled etching is also seen in the image shown earlier in Fig. 5 and is maintained for extended durations of etching, up to the removal of the dense surface layer, verifying that the etching is due to a change in chemistry within the porous domains. That is to say that when the dense surface layer is etched away, from the inside outwards, the unique chemical mixture that has built up within the domain is released and the etching is no longer enhanced by the etch products. This results in the cessation of etching, hinted at by the observation that bubbles stop emanating from the sample. It follows that the process is self-limiting and can therefore be used for highly reproducible formation of tetrahedral voids in the InP surface. It also follows that the amount of etching of the porous domain and dense surface layer can be controlled by controlling the duration of etching to any time up to times corresponding to the self-limiting case.

Increasing the temperature past 70°C results in an increase in the aggressiveness and lack of control of the etching resulting in highly disordered etching of the porous structures (as seen in Fig. 8b).

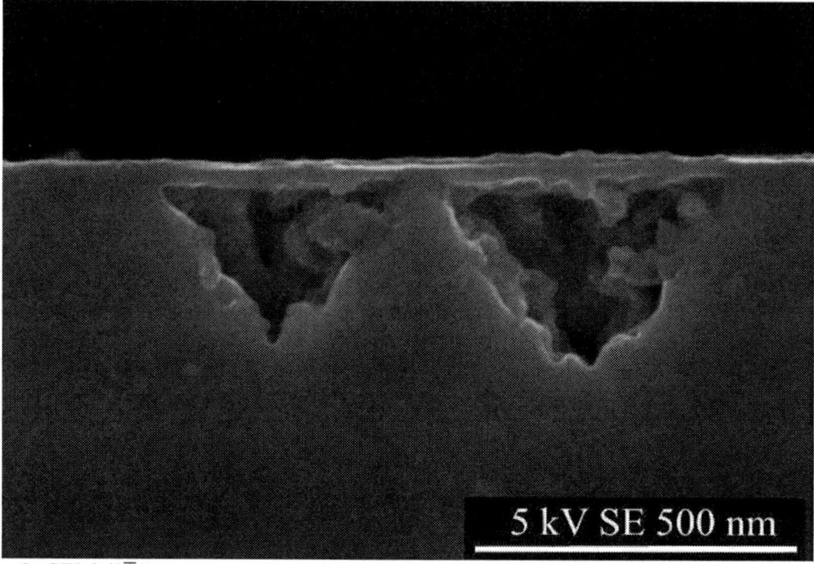

Fig. 9 SEM ($0\bar{1}1$) cross-section of porous domains in InP that have been etched in 7.5 mol dm^{-3} H$_2$SO$_4$ for 90 s at 70°C displaying and etched interior and an intact dense surface layer.

It can be clearly seen in Fig. 9 that there are regions of unetched semiconductor beneath the dense surface layer and outside of the original outlines of the domains of pores. Therefore many of the domains have not begun to merge together but due to the non-uniform nucleation and distribution of surface pits some domains begin to merge together much sooner than others. Fig. 10 shows an SEM cross-section of several domains from the same sample shown in Fig. 9. Due to the proximity of the surface pits of these domains, the domains have merged together and therefore the voids created during chemical etching overlap each other. It can be observed in the image that, although the triangular nature of the domain outlines can still be seen, the dense surface layer (at D) has been etched considerably resulting in its break-up. This type of etching, where the near-surface layer breaks up, only occurred where the domains were merged with each other. It can also be observed in this image that the etching has a slightly greater degree of disorder than in Fig. 9, even though they belong to the same experiment. Possible reasons for this include the trapping of products in the slightly narrower pores that form at the edge of merging domains. Any such trapping would result in magnification of the chain reaction, leading to more-enhanced, more-disordered etching. Furthermore, the structural weakness of a dense surface layer that spans more than one domain could result in the collapse of the suspended layer.

This type of overetching may be prevented by greater control of porous domain formation during anodisation. As has been noted in previous publications(31, 54), the pits that form at E_{pit} preferentially form at defects. Therefore the porous domains could be evenly distributed by patterning the InP surface with a regular distribution of defects using a focused ion beam (FIB) or by forming a regular array of etch pits on the surface using a photolithographically patterned etching process. In such a manner, controlled fabrication of complex devices such as wave guides or gas sensors could be realised in InP, using only a few simple chemical/electrochemical steps to produce sub-surface structures similar to those discussed above. Since anodisation results in similar pore formation in other semiconductors, this technique may be applicable to a range of materials including GaAs, InSb and GaP. Indeed, research on such materials may elucidate the underlying chemical mechanism behind the selective dissolution of the porous skeleton.

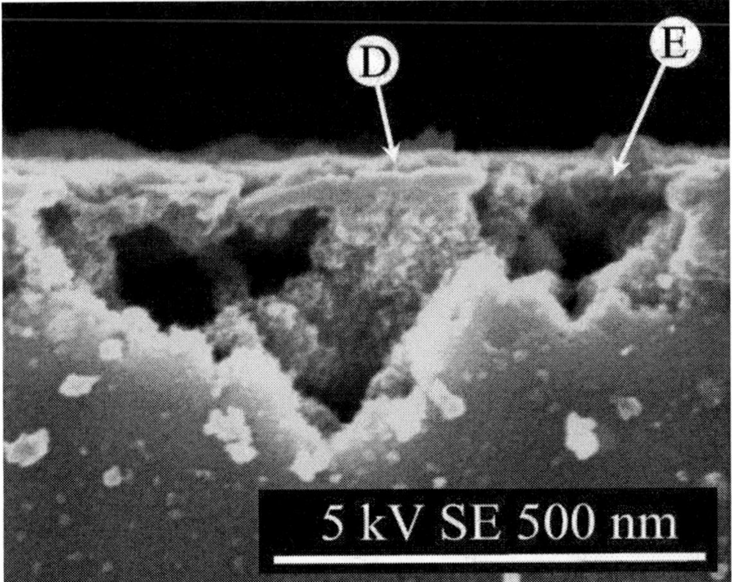

Fig. 10 SEM ($0\bar{1}1$) cross-section of InP containing porous domains that have been etched in 7.5 mol dm^{-3} H$_2$SO$_4$ for 90 s at 70°C. The cross-section image shows a damaged near-surface layer (at D) with some of the layer completely removed (at E).

Conclusions

During the early stages of electrochemical pore formation in InP in KOH, isolated domains of pores in the shape of truncated tetrahedra can be formed beneath a thin layer of dense (*i.e.* low porosity) InP. The porous structures formed can be chemically etched by 7.5 mol dm^{-3} H$_2$SO$_4$ at temperatures above 60°C. It follows that the process of etching these domains allows for the engineering of suspended 40-nm-thick layers above truncated tetrahedral voids. Greatest control of the etching process occurs at temperatures near 70°C and the degree of etching can be controlled by the etching time. The mechanism of the chemical etch seems to be dependent on the trapping of etch products within the porous structure resulting in the internal InP skeleton of the porous domains being etched by the H_2SO_4 at a higher rate than the surface layer of dense InP. It follows that once the dense surface layer that is containing these etch products is itself etched, the products diffuse away and the etching stops; *i.e.* the mechanism is self-limiting. It follows that this new technique can be used for the controlled formation of a range of structures including suspended nano-films and truncated tetrahedral voids. Such a technique may be a very useful tool in the fabrication of devices based on InP and it may be possible to extend the technique to the fabrication of free-standing InP nanofilms.

Acknowledgments

R. P. Lynch and N. Quill would like to thank the Irish Research Council (IRC) for PhD scholarships to perform this research. R. P. Lynch acknowledges a joint IRC - Marie Skłodowska Curie Fellowship under grant no. INSPIRE PCOFUND-GA-2008-229520. The authors would also like to acknowledge the support of the Tyndall National Institute through National Access Programme Funding.

References

1. R. L. Smith and S. D. Collins, *J. Appl. Phys.*, **71**, R1 (1992).
2. S. Langa, J. Carstensen, M. Christophersen, H. Foll and I. M. Tiginyanu, *Appl. Phys. Lett.*, **78**, 1074 (2001).
3. G. Oskam, A. Natarajan, P. C. Searson and F. M. Ross, *Appl. Surf. Sci.*, **119**, 160 (1997).
4. S. Langa, J. Carstensen, I. M. Tiginyanu, M. Christophersen and H. Foll, *Electrochem. Solid-State Lett.*, **4**, G50 (2001).
5. S. Langa, I. M. Tiginyanu, J. Carstensen, M. Christophersen and H. Foll, *Electrochem. Solid-State Lett.*, **3**, 514 (2000).
6. P. Schmuki, J. Fraser, C. M. Vitus, M. J. Graham and H. S. Isaacs, *J. Electrochem. Soc.*, **143**, 3316 (1996).
7. P. Schmuki, D. J. Lockwood, H. J. Labbe and J. W. Fraser, *Appl. Phys. Lett.*, **69**, 1620 (1996).
8. L. Santinacci, A.-M. Gonçalves, M. Bouttemy and A. Etcheberry, *J. Solid State Electrochem.*, **14**, 1177 (2010).
9. A. Eb, A.-M. Gonçalves, L. Santinacci, C. Mathieu and A. Etcheberry, *Comptes Rendus Chimie*, **11**, 1023 (2008).
10. L. Santinacci, M. Bouttemy, I. Gerard and A. Etcheberry, *ECS Trans.*, **19**, 313 (2009).
11. B. H. Erné, D. Vanmaekelbergh and J. J. Kelly, *Adv. Mater.*, **7**, 739 (1995).

12. J. G. Rivas, A. Lagendijk, R. W. Tjerkstra, D. Vanmaekelbergh and J. J. Kelly, *Appl. Phys. Lett.*, **80**, 4498 (2002).

13. M. Christophersen, J. Carstensen, A. Feuerhake and H. Föll, *Mat. Sci. Eng., B*, **69-70**, 194 (2000).

14. N. Quill, R. P. lynch, C. O'Dwyer and D. N. Buckley, *ECS Trans.*, **50**, 131 (2013).

15. N. Quill, R. P. Lynch, C. O'Dwyer and D. N. Buckley, *ECS Trans.*, **50**, 377 (2013).

16. A. M. Goncalves, L. Santinacci, A. Eb, I. Gerard, C. Mathieu and A. Etcheberry, *Electrochem. Solid-State Lett.*, **10**, D35 (2007).

17. R. P. Lynch, C. O'Dwyer, D. N. Buckley, D. Sutton and S. Newcomb, *ECS Trans.*, **2**, 131 (2006).

18. S. Ronnebeck, J. Carstensen, S. Ottow and H. Foll, *Electrochem. Solid-State Lett.*, **2**, 126 (1999).

19. P. Schmuki, L. E. Erickson, D. J. Lockwood, J. W. Fraser, G. Champion and H. J. Labbe, *Appl. Phys. Lett.*, **72**, 1039 (1998).

20. V. Lehmann and H. Foll, *J. Electrochem. Soc.*, **137**, 653 (1990).

21. T. Unagami, *J. Electrochem. Soc.*, **127**, 476 (1980).

22. J. Cartensen, M. Christophersen and H. Foll, *Mat. Sci. Eng., B*, **69-70**, 23 (2000).

23. F. M. Ross, G. Oskam, P. C. Searson, J. M. Macaulay and J. A. Liddle, *Philos. Mag. A*, **75**, 525 (1997).

24. M. J. J. Theunissen, *J. Electrochem. Soc.*, **119**, 351 (1972).

25. M. I. J. Beale, J. D. Benjamin, M. J. Uren, N. G. Chew and A. G. Cullis, *J. Cryst. Growth*, **73**, 622 (1985).

26. P. Allongue and C. H. de Villeneuve, *Appl. Phys. Lett.*, **67**, 941 (1995).

27. V. Lehmann and U. Gosele, *Appl. Phys. Lett.*, **58**, 856 (1991).

28. J.-N. Chazalviel, F. Ozanam, N. Gabouze, S. Fellah and R. B. Wehrspohn, *J. Electrochem. Soc.*, **149**, C511 (2002).

29. X. G. Zhang, *J. Electrochem. Soc.*, **138**, 3750 (1991).

30. X. G. Zhang, *J. Electrochem. Soc.*, **151**, C69 (2004).

31. R. P. Lynch, N. Quill, C. O'Dwyer, S. Nakahara and D. N. Buckley, *Phys. Chem. Chem. Phys.*, **15**, 15135 (2013).

32. L. Santinacci, I. Gerard, M. Bouttemy, M. Marques and A. Etcheberry, *ECS Trans.*, **16**, 411 (2008).

33. R. P. Lynch, C. O'Dwyer, N. Quill, S. Nakahara, S. B. Newcomb and D. Noel Buckley, *J. Electrochem. Soc.*, **160**, D260 (2013).

34. R. P. Lynch, C. O'Dwyer, D. Sutton, S. B. Newcomb and D. N. Buckley, *ECS Trans.*, **6**, 355 (2007).

35. C. O'Dwyer, D. N. Buckley, D. Sutton, M. Serantoni and S. B. Newcomb, *J. Electrochem. Soc.*, **154**, H78 (2007).

36. N. Quill, R. P. lynch, C. O'Dwyer and D. N. Buckley, *ECS Trans.*, **50**, 143 (2013).

37. C. O'Dwyer, D. N. Buckley, D. Sutton and S. B. Newcomb, *J. Electrochem. Soc.*, **153**, G1039 (2006).

38. R. P. Lynch, M. Dornhege, P. S. Bodega, H. H. Rotermund and D. N. Buckley, *ECS Trans.*, **6**, 331 (2007).

39. R. P. Lynch, N. Quill, C. O'Dwyer, M. Dornhege, H. H. Rotermund and D. N. Buckley, *ECS Trans.*, **53**, 65 (2013).
40. T. Sato, A. Mizohata, N. Yoshizawa and T. Hashizume, *Applied Physics Express*, **1**, 051202 (2008).
41. T. Sato, T. Fujino, A. Mizohata and T. Hashizume, in *E-MRS Fall Meeting 2007, Symposium B*, Warsaw (2007).
42. A. Salehi and D. Jamshidi Kalantari, *Sens. Actuators, B*, **122**, 69 (2007).
43. I. M. Tiginyanu, I. V. Kravetsky, S. Langa, G. Marowsky, J. Monecke and H. Föll, *Phys. Stat. Sol., A*, **197**, 549 (2003).
44. H. Tsuchiya, M. Hueppe, T. Djenizian, P. Schmuki and S. Fujimoto, *Sci. Tech. Adv. Mater*, **5**, 119 (2004).
45. G. Flamand and J. Poortmans, *Phys. Stat. Sol., A*, **202**, 1611 (2005).
46. H. Foll, J. Cartensen and S. Frey, *J. Nanometer.*, 1 (2006).
47. V. Kochergin and H. Foell, *Materials Science and Engineering: R: Reports*, **52**, 93 (2006).
48. A. Salehi, A. Nikfarjam and D. J. Kalantari, *Sens. Actuators, B*, **113**, 419 (2006).
49. E. H. M. Camara, C. Pijolat, J. Courbat, P. Breuil, D. Briand and N. F. de Rooij, *Transducers and Eurosensors '07* (2007).
50. J. H. Kwon, S. H. Lee and B. K. Ju, *J. Appl. Phys.*, **101**, 104515 (2007).
51. J. Mizsei, *Thin Solid Films*, **515**, 8310 (2007).
52. B. H. Erne, D. Vanmaekelbergh and J. J. Kelly, *J. Electrochem. Soc.*, **143**, 305 (1996).
53. R. W. Tjerkstra, J. Gomez Rivas, D. Vanmaekelbergh and J. J. Kelly, *Electrochem. Solid-State Lett.*, **5**, G32 (2002).
54. P. Schmuki, U. Schlierf, T. Herrmann and G. Champion, *Electrochim. Acta*, **48**, 1301 (2003).

Chapter 2

Compound Semiconductor Characterization

ECS Transactions, 69 (14) 51-57 (2015)
10.1149/06914.0051ecst ©The Electrochemical Society

Terahertz Spectroscopy: Studying Carrier Dynamics in Semiconductor Nanostructures

Lyubov V. Titova,[a,b] Sijia Xu,[b] Jean-Marc Baribeau,[c] David J. Lockwood,[c] and Frank A. Hegmann[b]

[a] Department of Physics, Worcester Polytechnic Institute, Worcester, MA 01609, USA
[b] Department of Physics, University of Alberta, Edmonton, Alberta T6G 2E1, Canada
[c] National Research Council, Ottawa, ON K1A 0R6, Canada

Understanding the ultrafast dynamics of photoexcited carriers in semiconductor nanostructures and their dependence on sample morphology is crucial for their incorporation into photonic devices. Time-resolved terahertz (THz) spectroscopy (TRTS) is an all-optical, contact-free technique that directly measures the transient mobile carrier dynamics and terahertz conductivity in materials over picosecond time scales, and is uniquely suited as a probe of conductivity in nanomaterials. Using low temperature MBE-grown silicon films as an example, we show how TRTS can be used to probe microscopic photoconductivity as well as obtain crucial insights into sample morphology. The thin silicon films consist of a mixture of amorphous and crystalline phases, and their relative content changes drastically with growth temperature. Photoexcited carrier dynamics in these films are determined by film crystallinity: in the amorphous phase, carriers are trapped in bandtail states on sub-picosecond time scales, while the carriers excited in crystalline grains remain free for tens of picoseconds. The complex THz conductivity spectra obtained from the TRTS measurements show that the long range conductivity is significantly higher in films grown at higher temperatures that contain a larger fraction of crystalline material with larger crystal grain sizes.

Nanostructured and granular materials are often easier and less expensive to produce compared to single crystals, and are thus actively explored as alternatives to single crystalline materials for photonic applications, and in particular for low-cost thin film solar energy conversion devices (1,2).

Understanding the dynamics of optically excited carriers in these materials over length scales of the grain size and over picosecond time scales is necessary for integrating them into new photonic devices. Time-resolved terahertz (THz) spectroscopy (TRTS) is a non-contact, all-optical probe of conductivity dynamics on ultrafast time scales following optical injection and over nanometer length scales that is uniquely suited for exploring carrier dynamics in nanostructured materials (3-17). Correlating time-resolved THz photoconductivity and film morphology as determined by electron microscopy, x-ray diffraction or other structural characterization techniques, yields information about microscopic conductivity and photoexcited carrier lifetimes that is essential for designing high-efficiency photovoltaic devices. We have applied TRTS to investigate ultrafast carrier dynamics in a variety of nanostructured semiconducting systems such as silicon nanocrystal films (3, 16), nanogranular iron pyrite (FeS_2,22) and vanadium dioxide (VO_2,

4). Furthermore, terahertz emission spectroscopy of nanomaterials, where the sample itself emits THz radiation in response to optical excitation, provides important insights into carrier dynamics, and has been applied to study a variety of nanostructured materials such as InAs and GaAs nanowires (23-25) and aligned single-wall carbon nanotube arrays (26).

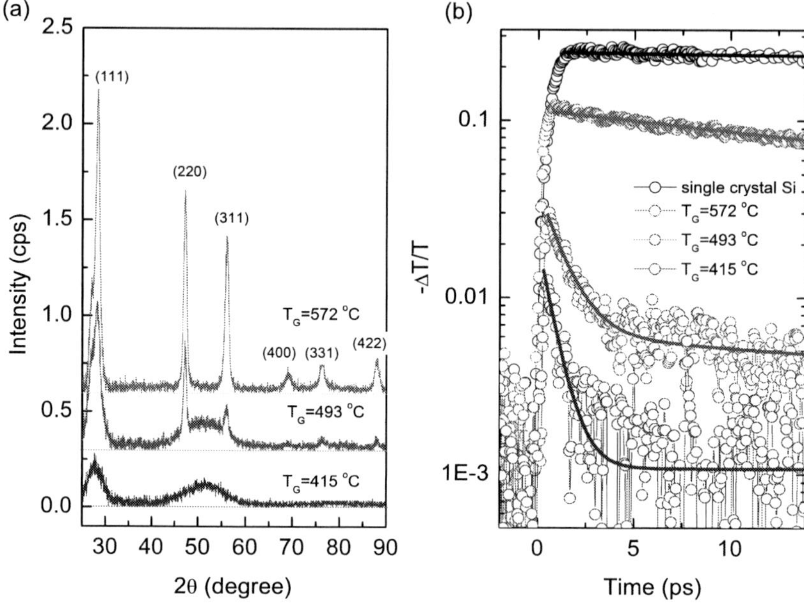

Figure 1. a) Evolution of the glancing angle XRD profile from thin Si films on fused quartz as a function of growth temperature. (b) Change in transmission of the main peak of the THz probe pulse, $-\Delta T/T$, as a function of time delay with respect to a 400 nm, 100 fs pump pulse for single crystal Si on sapphire and for Si films grown at different temperatures, measured for a pump fluence of 220 $\mu J/cm^2$. Solid lines are fits to either a double-exponential decay (growth temperature 493 °C) or a single-exponential decay (all other samples).

Here we use TRTS to study photoexcited carrier dynamics in pure, non-hydrogenated Si thin films with varying degree of crystallinity grown by MBE on quartz substrates at temperatures ranging from 415 °C to 572 °C. Low temperature MBE-grown thin amorphous and nanocrystalline Si films have many advantages over conventionally grown amorphous and polycrystalline Si layers for photovoltaic applications: they have 98% of the mass density of crystalline Si and exceptionally low (i.e., 0.1-1 atomic %) hydrogen content, compared to over 10 atomic % in conventionally grown layers (27, 28). Correlating film microstructure as determined by XRD and TEM measurements with photoexcited conductivity dynamics over picosecond time scales after excitation, we find that the photoexcited carrier lifetimes are determined by the film crystallinity, since bandtail states in amorphous Si rapidly trap photoexcited carriers. Applying the Drude-

Smith model to analyze the complex photoconductivity of the films, we find that the localization that photoexcited carriers experience decreases with increasing growth temperature, and estimate crystal grain sizes to be ~ 4 nm in a film grown at 493 °C and ~ 12 nm in a film grown at 572 °C.

The Si films were grown on fused quartz substrates using a VG Semicon V80 UHV MBE system at a variety of different growth temperatures, T_g, using a fixed deposition rate of 0.2 nm/s. The thickness of the films ranged from 170 nm to 190 nm. The volume fraction of the crystalline phase increases steeply with increase in growth temperature, from 8% in films grown at 415 °C to 98% in films grown at 572 °C, as determined from Raman spectroscopy, XRD and TEM analysis (27,29). Figure 1(a) shows grazing incidence x-ray diffraction (GIXRD) from various low-temperature-grown Si thin films obtained at an angle of incidence α = 0.3°. In the film grown at 415 °C, the amorphous bands at 27° and 52° are evident. At T_g=572 °C, the amorphous bands are extremely weak, while the crystalline diffraction peaks are intense, indicating that polycrystalline Si is the dominant phase. At T_g= 493 °C, polycrystalline diffraction lines are superimposed on the amorphous Si bands, indicating that the film contains a mixture of the two phases, amorphous and crystalline. XRD results are in good agreement with TEM measurements performed on analogous films grown by MBE at the same growth temperatures on Si substrates covered with a thin native oxide Si (27). TEM measurements show that the film grown at T_g=572 °C is mostly polycrystalline, with 10-20 nm thick needles of single crystals interspersed with various chunks of polycrystalline material exhibiting twinning and other defects. In the thin layers at the top of the sample as well as at the film-substrate interface, small Si crystallites are intermixed with amorphous Si. In the film grown at T_g= 493 °C, smaller crystallites are surrounded by amorphous Si, and the film grown at T_g= 415 °C is entirely amorphous.

Correlating this information regarding the microstructure of the films with THz photoconductivity provides important information about structure-function relations in low temperature MBE grown Si films. The TRTS technique and its application to probe photoconductivity have been described in detail previously (3,4). In brief, charge carriers are optically injected into the sample with an ultrafast optical pump pulse. In the experiments described here, 400 nm, 100 fs pump pulses were used. A time-delayed THz pulse is used to probe pump-induced changes in the sample, as free-carrier absorption occurs in semiconductors in the THz spectral range. THz probe pulses with a bandwidth spanning frequencies from 0.4 to 2.5 THz (or 1.6 to 10.3 meV) were generated by optical rectification of 800 nm, 100 fs pulses from a 1 kHz amplified Ti:sapphire laser source in a [110] ZnTe crystal, and coherently detected by free-space electro-optic sampling in a second [110] ZnTe crystal. TRTS probes free carrier motion over length scales $L(\omega) \propto \sqrt{D/\omega}$, where D is the diffusion constant, and ω is the THz probe frequency, corresponding to a length scale of ~ 10-20 nm in Si for a THz pulse bandwidth of 0.4 – 2.5 THz. Varying the delay between the pump pulse and THz probe pulse, we track pump-induced changes in the sample properties with sub-picosecond time resolution by monitoring the change $\Delta T(t)$ in transmission of the main peak of the THz probe pulse. In the limit of small transmission modulation, $-\Delta T(t)$ is proportional to the time-dependent photoconductivity (3,7).

The transient THz conductivity of three MBE-grown Si films, as well as a crystalline epitaxial Si film on sapphire, excited by 220 μJ/cm^2 pulses is shown in Fig.1 (b). Unlike the crystalline epitaxial Si sample that exhibits long-lived photoconductivity, the photoconductivity signal in the MBE-grown Si films is followed by a fast decay due to carrier trapping and/or electron-hole recombination (30). The THz photoconductivity dynamics depends on the crystallinity of MBE-grown Si films. For the amorphous film grown at T_g=415 °C, the photoinduced conductivity rapidly decays to zero on a sub-picosecond time scale, as has been previously observed in amorphous Si (6,30-32). The single-exponential decay fit to the conductivity dynamics of this film (solid lines) yields a carrier lifetime of 0.8 ± 0.1 ps. The photoconductivity decay of the polycrystalline film (T_g=572 °C) has a peak value comparable to that of single crystal Si on sapphire. It also follows a single-exponential decay with a carrier trapping time of 34 ps. The film grown at 493 °C, which, according to TEM and XRD measurements, consists of small polycrystalline inclusions embedded in an amorphous silicon matrix (29), exhibits the double-exponential photoconductivity decay, with a fast (~ 1ps) component representing fast carrier trapping in the amorphous phase, and a longer (29 ps) decay associated with carriers in crystalline grains. This correspondence between THz photoconductivity dynamics and microstructural analysis suggests that TRTS can serve as a rapid characterization tool for new nanostructured materials for photovoltaic and other photonic devices.

Even more detailed information about microscopic photoconductivity and carrier localization can be obtained by analyzing the complex THz photoconductivity at fixed time intervals after optical injection. Frequency-dependent THz photoconductivity at a given time point following the photoexcitation can be measured by fixing the delay between the pump pulse and THz probe pulse, and detecting the pump-induced changes in the amplitude and phase of the THz pulse transmitted through the sample.

Figure 2 shows the real and imaginary components of the THz conductivity spectra of the single-crystal epitaxial Si film and the polycrystalline Si film grown at 572 °C and 493 °C taken 3 ps after excitation with 400 nm, 100 fs pump pulses at a fluence of 220 μJ/cm^2. The single crystal Si on sapphire thin film (Fig. 2(a)) exhibits Drude-like conductivity with carrier scattering time τ_D=39±1 fs. The photoconductivity spectra of both low temperature MBE grown films (Fig. 2 (b) and (c)) appear markedly different from those of a single crystalline Si: imaginary conductivity is negative and the low-frequency real conductivity is suppressed, which is typically observed in materials where mobile carriers are localized within crystallites and nanoscale grains. We analyze the observed THz conductivity spectra within the framework of the Drude-Smith model, a modified the Drude model that takes into account the effects of charge carrier confinement over nanometer length scales (3-5, 8-15, 33, 34). In this model, the complex conductivity as given by

$$\hat{\sigma}(\omega) \;=\; \frac{\varepsilon_0 \omega_p^2 \tau_{DS}}{1 - i\omega\tau_{DS}}\left[1 + \frac{c}{1 - i\omega\tau_{DS}}\right], \qquad\qquad [1]$$

where $\omega_p^2 = ne^2 / \varepsilon_0 m^*$ is the plasma frequency, n is the charge carrier density, m^* is the carrier effective mass and τ_{DS} is the effective scattering time. Solid lines in Fig. 2 (b) and

(c) are simultaneous fits of real and imaginary components of photoconductivity in films grown at T_g=572 °C (Fig. 2(b)) and T_g=493 °C (Fig. 2(c)).

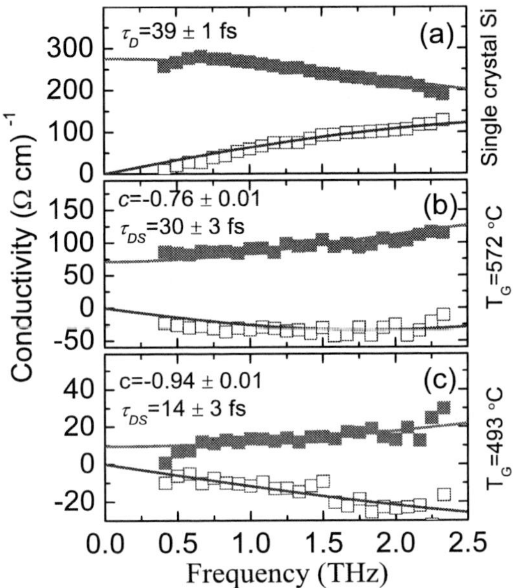

Figure 2. Real (solid red squares) and imaginary (open blue squares) components of the complex conductivity of photoexcited single crystal Si on sapphire (a), and of MBE grown Si films with T_g=572 °C (b) and T_g=493 °C (c). Spectra in (a) were measured 5 ps after photoexcitation, and spectra in (b) and (c) - 3 ps after photoexcitation with 400 nm pump pulses with the pump fluence of 220 µJ/cm^2. Lines are fits to the Drude (a) or Drude-Smith model (Eq. [1]) (b, c).

The parameter c represents the degree of carrier localization, where c = -1 describes the situation where free carriers are completely localized within crystalline grains with dimensions on the order of the bulk mean free path, and c = 0 corresponds to free carrier motion according to the Drude model. The scattering time τ_{DS} is also modified from the bulk value τ_D in order to take into account the reduction of the carrier mean free path by nanocrystal or grain boundaries (3, 5). Carriers experience significant localization within crystalline grains in both low temperature grown films, however, and effects of localization are much more pronounced in the film grown at lower temperature T_g=493 °C, with c=-0.94, compared to c=-0.76 for the film with T_g=572 °C. Extrapolating the fit of the real conductivity component to zero frequency, we can also determine σ_{DC}, the dc photoconductivity at a given time after photoexcitation. As expected, we find that σ_{DC} is significantly higher in the more crystalline film grown at T_g=572 °C compared to the film grown at T_g=493 °C (~ 70 (Ω cm)$^{-1}$ versus ~ 10 (Ω cm)$^{-1}$ at 3 ps following photoexcitation, respectively). For both low-temperature-grown films, however, σ_{DC} is much lower than that in the single crystal Si on sapphire sample (~ 275 $(\Omega\,\text{cm})^{-1}$) at the same photoexcitation fluence of 220 $\mu J/cm^2$.

Analyzing the THz photoconductivity spectra also allows us to estimate the size of crystalline grains in the two films. The effective scattering time τ_{DS} is related to the bulk scattering time and the average time between collisions with the grain boundaries $\tau_{boundary}$ according to Matthiessen's Rule, $1/\tau_{DS} = 1/\tau_{bulk} + 1/\tau_{boundary}$ (3,5). Taking the bulk Si scattering time for the single crystal Si on sapphire as bulk scattering time $\tau_{bulk}=$ 39±1 fs (Fig. 2(a)), we can use τ_{DS} to estimate average crystalline grain size as $d = v_{th}\dfrac{\tau_{bulk}\tau_{DS}}{\tau_{bulk}-\tau_{DS}}$, where $v_{th} = 2*10^5$ m/s is the thermal velocity for electrons in silicon at room temperature. We find the grain size to be ~ 26 nm for the film grown at T_g=572 °C, and ~ 4 nm for the film grown at T_g=493 °C, in good agreement with TEM measurements (29).

In summary, we have applied time-resolved THz spectroscopy measurements to study thin pure, non-hydrogenated MBE-grow Si films. We find that carrier lifetimes and photoconductivity are determined by film crystallinity, and that TRTS measurements not only provide insight into photoexcited carrier dynamics but also can be used to obtain crucial information about film morphology, such as the degree of crystallinity and average sizes of crystalline grains in polycrystalline films. TRTS is thus uniquely suited for rapid, contact-free characterization of new nanostructured materials for photovoltaics and other photonic devices.

Acknowledgments

We thank G. Popowich and D. Mullin for technical assistance and X. Wu for the TEM work on related Si/SiO$_2$ samples. We acknowledge support from NSERC, CFI, ASRIP, and AITF iCiNano.

References

1. P.V. Kamat, *J. Phys. Chem. C*, **111**, 2834 (2007).
2. M. Beard, J.M. Luther, and A.J. Nozik, *Nat. Nanotech.*, **9**, 951 (2014).
3. L. V. Titova, T. L. Cocker, D. G. Cooke, X. Wang, A. Meldrum, and F. A. Hegmann, *Phys. Rev. B*, **83**, 085403 (2011).
4. T. L. Cocker, L. V. Titova, S. Fourmaux, G Holloway, H. -C. Bandulet, D. Brassard, J. -C. Kieffer, M. A. El Khakani, and F. A. Hegmann, *Phys. Rev. B*, **85**, 155120 (2012).
5. G.M. Turner, M.C. Beard, and C.A. Schmuttenmaer, *J. Phys. Chem. B*, **106**, 11716 (2002).
6. L. Fekete, P. Kužel, H. Němec, F. Kadlec, A. Dejneka, J. Stuchlík, and A. Fejfar, *Phys. Rev. B*, **79**, 115306 (2009).
7. D. G. Cooke, A. N. MacDonald, A. Hryciw, J. Wang, Q. Li, A. Meldrum, and F. A. Hegmann, *Phys. Rev. B*, **73**, 193311 (2006).
8. M. Beard, G. Turner, J. E. Murphy, O. I. Micic, M. C. Hanna, A. J. Nozik, and C. Schmuttenmaer, *Nano Lett.*, **3**, 1695 (2003).

9. M. Walther, D. G. Cooke, C. Sherstan, M. Hajar, M. R. Freeman, and F. A. Hegmann, *Phys. Rev. B,* **76**, 125408 (2007).
10. R. Lovrinčić and A. Pucci, *Phys. Rev. B,* **80**, 205404 (2009).
11. H. Ahn, Y.-P. Ku, Y.-C. Wang, C.-H. Chuang, S. Gwo, and C.-L. Pan, *Appl. Phys. Lett.,* **91**, 163105 (2007).
12. J. B. Baxter, and C. A. Schmuttenmaer, *J. Phys. Chem. B, 110,* 25229 (2006).
13. P.D. Cunningham, L.M., Hayden, H.-L. Yip, A.K.-Y. Jen, *J. Phys. Chem B,* **113**, 15427 (2009).
14. Q.-I. Zhou, Y. Shi, B. Jin, and C. Zhang, *Appl. Phys. Lett.,* **93**, 102103 (2008).
15. D. Tsokkou, A. Othonos, and M. Zervos, *Appl. Phys. Lett.,* **100**, 133101 (2012).
16. C.J. Docherty, P. Parkinson, H.J. Joyce, M.-H. Chiu, C.-H. Chen, M.-Y. Lee, L.-J. Li, L.M. Herz, and M.B. Johnston, *ACS Nano,* **8**, 11147 (2014).
17. L. V. Titova, T. L. Cocker, X. Wang, A. Meldrum, and F. A. Hegmann, *ECS Trans.,* **45**, 21 (2012).
18. P.U. Jepsen, D.G. Cooke, and M. Koch, *Laser Photon. Rev.,* **5**, 124 (2011).
19. J.B. Baxter, G.W. Guglietta, *Anal. Chem.,* **83**, 4342 (2011).
20. H.J Joyce, C.J. Docherty, Q. Gao, H. H. Tan, C. Jagadish, J. Lloyd-Hughes, L.M. Herz and M.B. Johnston, *Nanotechnology,* **24**, 214006 (2013).
21. J.B. Baxter and C.A. Schmuttenmaer, *J. Phys. Chem. B,* **110**, 25229 (2006).
22. J.M. LaForge, B. Gyenes, S. Xu, L.K. Haynes, L.V. Titova, F.A. Hegmann, M.J. Brett, *Sol. Energ. Mat. Sol. Cells,* **117**, 306 (2013).
23. D.V. Seletskiy, M.P. Hasselbeck, J.G. Cederberg, A. Katzenmeyer, M.E. Toimil-Molares, F. Léonard, A. A. Talin, and M. Sheik-Bahae, *Phys. Rev. B,* **84**, 115421 (2011).
24. V. N. Trukhin, A.S. Buyskikh, N.A. Kaliteevskaya, A.D. Bourauleuv, L.L. Samoilov, Y.B. Samsonenko, G.E. Cirlin, M.A. Kaliteevski, and A.J. Gallant, *Appl. Phys. Lett.,* **3**, 072108 (2013).
25. A. Arlauskas, J. Treu, K. Saller, I. Beleckaitė, G. Koblmüller, and A. Krotkus, *Nano Lett.,* **14**, 1508 (2014).
26. L.V. Titova, C.L. Pint, Q. Zhang, R.H. Hauge, J. Kono, and F. A. Hegmann, *Nano Lett.,* **15**, 3267 (2015).
27. J.-M. Baribeau, X. Wu, D.J. Lockwood, L. Tay, and G.I. Sproule, *J. Vac. Sci. Technol B,* **22**, 1479 (2004).
28. B. J. Fogal, S. K. O'Leary, D. J. Lockwood, J.-M. Baribeau, M. Noël, and J. C. Zwinkels, *Solid State Commun.,* **120**, 429 (2001).
29. A. Akbari-Sharbaf, J.-M. Baribeau, X. Wu, D.J. Lockwood, and G. Fanchini, *Thin Solid Films,* **527**, 38 (2013).
30. A. Esser, K. Seibert, H. Kurz, G.N. Parsons, C. Wang, B.N. Davidson, G. Lucovsky, and R.J. Nemanich, *Phys. Rev. B* **41**, 2880 (1990).
31. K.E. Myers, Q. Wang, and S.L. Dexheimer, *Phys. Rev. B,* **64**, 161309(R) (2001).
32. K.P.H. Lui and F.A. Hegmann, *J. Appl. Phys.,* **93**, 9012 (2003).
33. N.V. Smith, *Phys. Rev. B,* **64**, 155106 (2001).
34. T. L. Cocker, L. V. Titova, S. Fourmaux, H. -C. Bandulet, D. Brassard, J. -C. Kieffer, M. A. El Khakani, and F. A. Hegmann, *Appl. Phys. Lett.,* **97**, 221905 (2010).

Chapter 3

Optoelectronics

ECS Transactions, 69 (14) 61-75 (2015)
10.1149/06914.0061ecst ©The Electrochemical Society

Recent Developments in Mercury Cadmium Telluride IR Detector Technology

J. Antoszewski, N.D. Akhavan, G. Umana-Membreno, R. Gu, W. Lei, L. Faraone

School of Electrical, Electronic and Computer Engineering, The University of Western Australia, Western Australia 6009, Australia

In spite of many attempts to find better material for infrared detectors, HgCdTe still dominates the high performance end of the market. At present, the research in this field is directed towards high pixel density, high yield, reduced cooling and hyperspectral operation. Long lasting issue of availability of high quality, large area substrates for epitaxial growth of HgCdTe seems to be easing recently with development of 4" CdZnTe lattice matching substrates reported recently. The quality of HgCdTe grown on alternative substrates like silicon, GaAs or recently investigated GaSb still falls behind that achievable on CdZnTe due to large lattice mismatch leading to high defect density. Recent effort to scale down the pixel size and pitch resulted in development of diffraction limited high definition focal plane arrays. Implementation of unipolar n-type/barrier/n-type detector structure in HgCdTe material system has been recently proposed and studied intensively. Due to superior transport and optical properties of HgCdTe it is expected that, when optimized, these structures will allow for background limited performance at significantly higher operating temperatures.

Introduction

Since its invention in 1959, HgCdTe and associated with this material infrared detector technology evolved from bulk crystal growth and basic single photo-detectors to very sophisticated, epitaxially grown material and high density imaging focal plane arrays (FPA). However, present HgCdTe based IR systems still require efficient cryogenic cooling which is bulky and expensive. Consequently, the research and development in this field is directed towards high pixel density, reduced cooling, and high process yield. In addition, HgCdTe based multiband and hyperspectral IR systems are being investigated and developed. In order to meet these goals there is an urgent need for still better quality HgCdTe material and novel device architectures alleviating the factors limiting the performance of present day HgCdTe detectors.

In terms of HgCdTe material, the most challenging is development of large area, high quality epi-ready substrates for large size, high pixel density HgCdTe FPAs. Traditionally, lattice matched single crystal CdZnTe wafers have been used. However, the cost of these substrates is still high although size is continuously increasing. The readily available large size Si and GaAs substrates have been investigated for quite some time but due to large mismatch of lattice constant and coefficient of thermal expansion (CTE) the best reported dislocation densities in HgCdTe layers grown on these substrates are still about two orders of magnitude higher than in HgCdTe epi-layers grown on lattice matched CdZnTe. Recently better lattice matching GaSb epi-ready substrates become commercially available and are being investigated as potential replacement for CdZnTe.

The pixel size and pitch in the FPAs has already approached the diffraction limit (size of pixel comparable with IR wavelength). Further reduction of array pitch does not translate into better image resolution but reduction of FPA should result in better uniformity and higher yield.

Regarding device architectures, so called barrier structures attracted much attention within HgCdTe research community recently. The device based on the structure of n-type absorber/un-doped barrier/thin n-type contact (nBn), at least theoretically does not have built in voltage region (hence reduced dark current and noise) and if externally biased allows for free flow of minority photo-holes in valence band while blocking majority photo-electrons in conduction band. Implementation of this concept into HgCdTe material system is not straightforward due to valence band offset appearing at the n-type/barrier interfaces. However, this undesirable effect may be compensated by superior HgCdTe electron mobility and photo-generated carrier lifetime. Moreover, promising concepts for elimination of the valence band barrier in HgCdTe nBn structure have been proposed and will be discussed later in this paper.

Although HgCdTe was invented more than 50 years ago, its technology is still complex and yield relatively low, particularly for large FPA demanding over 99% pixel operability. Consequently, there is continuing search for more technologically forgiving replacement. Well established III-V technology always seemed to be an attractive option. The quantum well and quantum dots based FPA technologies have been developed but it appears that they are not able to match HgCdTe performance due to their inferior fundamental properties. Recently, type-II superlattices based on InAs/GaSb or InAs/InAsSb are heavily investigated. The barrier detectors based on these materials are expected to match or even outperform HgCdTe devices. However, it has not been demonstrated yet and there are indications that the inherently short Shockley-Reed-Hall (SRH) lifetime < 100ns in these materials may be a major obstacle. It remains to be seen if this problem can be solved in the future. In the meantime HgCdTe remains unchallenged with good prospects of achieving future goal of large FPA operating at room temperature.

SUBSTRATES

Although HgCdTe growth on CdZnTe lattice matched substrates provides the best quality material, CdZnTe is expensive and until recently of limited size. Therefore, over the past two decades, several alternative substrates have been studied to replace CdZnTe such as Si, Ge and GaAs [1-5]. All these substrates represent very high crystal quality, large wafer size, low cost, and ready availability. Among them, Si has attracted more attention due to its compatibility with the Si readout integrated circuit (ROIC) in a flip-chip bonded configuration. However, all these alternative substrates present a very large lattice and coefficient of thermal expansion (CTE) mismatch with HgCdTe as shown in Figure 1 [6]. This large lattice and CTE mismatch leads to a high defect density in HgCdTe epilayers, which not only reduces the carrier mobility and lifetime (and thus degrades detector performance), but also leads to a high density of defective/dead pixels in the FPA. A common approach to reducing dislocations in the HgCdTe layers is to grow a thick CdTe buffer layer in order to relax and block dislocations propagating from the lattice mismatched substrate/CdTe interface [2,4,7]. Figure 2 shows the dependence of full width at half maximum (FWHM) of the X-ray diffraction (XRD) peak of the CdTe buffer layer as a function of the buffer layer thickness and etch pit density (EPD) [4]. The CdTe buffer with an appropriate thickness (>10µm) can effectively block the penetration of dislocations, leading to CdTe buffer with XRD FWHM around 60 arc sec and dislocations density in low $10^6 cm^{-2}$ range, about two orders of magnitude higher that that achievable on CdZnTe. In case of Si substrates, a ZnTe nucleation layer and an As passivation layer are applied to suppress the formation of micro-twin defects in the buffer and subsequently grown HgCdTe epilayer [4,8]. In addition, other technologies reducing dislocation density have been studied including thermal cycling or dislocation gettering.

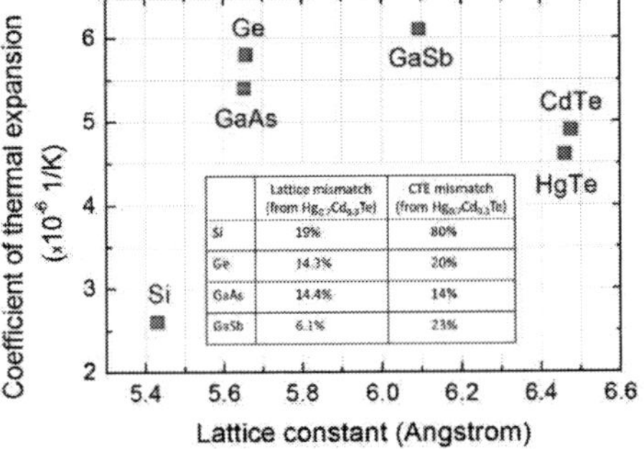

Fig.1 Room temperature lattice constant and CTEs of several semiconductors used as substrates for HgCdTe epitaxial growth. The inset shows the lattice and CTE mismatch between four potential alternative substrates and HgCdTe (after Ref.6).

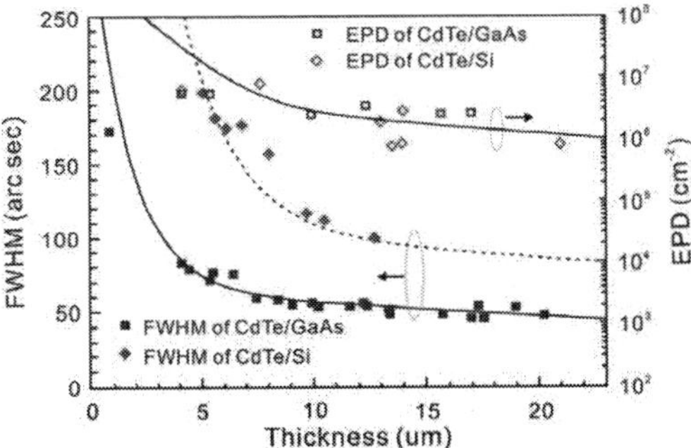

Fig. 2. XRD FWHM and EPD values for CdTe buffer layers grown on GaAs and Si in relation to CdTe buffer layer thickness (after Ref. 4).

Thermal cycle annealing reduces the dislocation density in the CdTe buffer and HgCdTe layers [7, 9-11]. The dislocation density in HgCdTe grown on CdTe/GaAs can be reduced from 7.7×10^6 cm^{-2} to as low as 2.3×10^5 cm^{-2} by thermal cycling between 300 and 490°C [11]. Although this process is very effective it leads to interdiffusion of elements degrading the sharpness of hetero- and homojunctions. Reticulated structures can be used to getter dislocations and hence reduce the dislocation density [12-14]. Stoltz et al. reported HgCdTe mesa structures (on CdTe/Si) with near-vertical sidewalls obtained with plasma etching [5]. Subsequent four thermal cycle anneals between 494°C and 250°C enabled dislocations to glide to sidewall stress-free regions resulting in reduction of dislocations within the device active areas from 1.1×10^7 cm^{-2} to 3.35×10^5 cm^{-2}, an order higher than achievable on CdZnTe.

Intensive study on alternative substrates over the past three decades provided significant advance in growth of HgCdTe on Si, Ge and GaAs alternative substrates. Table I shows summary of achieved material quality [4,15]. Although a substantial gap exists between CdTe and HgCdTe quality grown on alternative substrates compared to growth on CdZnTe, these substrates have been successfully used for production of MWIR HgCdTe detectors and FPAs with performance at 77K comparable to those grown on CdZnTe. However, the performance of LWIR HgCdTe detectors, particularly FPAs, grown on alternative substrates is still much lower than those on CdZnTe.

Table I Typical XRD FWHM and EPD values for CdTe and HgCdTe epilayers grown on CdZnTe, Si and GaAs substrates reported in Refs. 4 and 15.

As-grown (211) B	CdZnTe substrate	HgCdTe on CdZnTe	CdTe on Si	HgCdTe On Si	CdTe on GaAs	HgCdTe on GaAs
XRD FWHM (arc sec)	8	12	50	80	30	80
EPD (cm^{-2})	5×10^4	3×10^4	5×10^6	4×10^6	2×10^6	2×10^6

GaSb has been proposed recently as a new alternative given that epi-ready high quality substrates are now readily available in large wafer sizes [6]. With much smaller lattice and CTE mismatch, in principle GaSb provides a better choice as an alternative substrate. Table II compares the main characteristics of commercially available GaAs and GaSb substrates, indicating that GaSb substrates have similar IR transmission and commercial availability to GaAs substrates. In addition, GaSb substrates show better material quality in terms of XRD FWHM and EPD. Although the cost of GaSb substrates is higher, ultimately a smaller lattice mismatch should result in growth of HgCdTe with higher crystal quality and drive down cost and yield of large format LWIR FPAs.

Table II Main characteristics of GaAs and GaSb as alternative substrates for growing HgCdTe (after Ref.6)

(211)B wafer	XRD FWHM (arc sec) [*1]	EPD (cm^{-2}) [*2]	IR transmission (%) [*3]	Max. diam. (inch) [*4]	Unit price (US$ for 2″ wafer)
GaAs	24 ~ 30	<5000	48 ~55	6	~ 150
GaSb	20 ~ 25	<3000	47 ~ 52	4	~ 550

*1 For omega scan (according to Wafer technology, UK)
*1 From Wafer technology, UK.
*3 For 2~15 um wavelength range (according to Wafer technology, UK)
*4 Some manufacturers can provide even large size wafers.

Table III presents some characteristics of CdTe buffer layers grown on GaSb and GaAs substrates at University of Western Australia (UWA) [6]. Note that main characteristics of both materials are comparable and matching state-of-the-art results reported previously for CdTe on GaAs [3,16,17]. Typical characteristics of the initial non-optimized HgCdTe

Table III Selected material characteristics for MBE-grown CdTe buffer layers on GaAs and GaSb substrates at UWA (after Ref.6)

alternative substrate	CdTe buffer layer thickness (um)	XRD FWHM (arc sec)	RHEED pattern during CdTe buffer layer growth	Etch pit density ($\times 10^6$ cm^{-2})
GaAs	6 ~ 7	60 ~ 75	long and uniform streaks	3 ~ 50
GaSb	3 ~ 7	55 ~ 71	long and uniform streaks	5 ~ 30

epilayers grown on the reported above CdTe/GaSb substrates are compared with those grown on CdTe/GaAs in Table IV [18]. In terms of reflection high-energy electron diffraction (RHEED) pattern, XRD FWHM, and EPD, they demonstrate similar figures. It is expected that optimization of the growth process will improve the HgCdTe crystal quality.

Table IV Characteristics for MBE-grown HgCdTe layers on CdTe/GaAs and CdTe/GaSb substrates at UWA (after Ref.18)

alternative substrate	x value of Hg$_{1-x}$Cd$_x$Te layer	HgCdTe layer thickness (um)	CdTe buffer layer thickness (um)	XRD FWHM (arc sec)	RHEED pattern during HgCdTe layer growth	Etch pit density ($\times 10^6$ cm^{-2})
GaSb	0.27~0.32	4.5~5.3	5.3~5.7	122~139	long and uniform streaks	2~10
GaAs	0.26~0.32	5.7~6.7	5.7~6.7	98~155	long and uniform streaks	8~40

All above results indicate that from early 2000 the progress in the HgCdTe growth on alternative substrates stumbled on the unbreakable so far barrier associated with high level of threading dislocations. Although quality of HgCdTe layers grown on alternative substrates is sufficient for manufacturing of cryogenically cooled FPA, their properties are not meeting requirements for HgCdTe growth for high operating temperatures (HOT) conditions.

In the meantime, the recent progress in the growth of CdZnTe substrates has been significant. Mackenzie et al. (Redlen Technologies) reported CdZnTe single crystal material grown by traveling heater method (THM) rather than commonly used growth from melt (19). Their growth process has been shown to deliver routinely 20 wafers of 90mm x 90mm size from a single 100mm-diameter boule with uniform optical, electrical, and composition properties. The typical dislocation density was shown to be below 4×10^4cm^2 and FWHM of DXRC values approaching 40arcsec. Successful growth of even larger, 115mm-diameter (over 4") CdZnTe boules have been reported by researchers from LETI with wafers showing FWHM of DXRC in the range of 20-40arcsec and low 10^4cm^2 dislocations density (20). These achievements indicate that even larger diameter

CdZnTe substrates may be available in the near future, representing significant breakthrough in the supply of lattice matched, epi-ready substrates for HgCdTe growth.

PIXEL SIZE REDUCTION

The alternative for expanding substrate size is reduction of pixel size and pitch in a FPA. With smaller pixel area, more FPAs can be processed out of a single HgCdTe wafer, thus also reducing the FPA cost. Over the last two decades the HgCdTe available detector pixel size has decreased systematically from over 50μm to 30μm and then in the past several years to around 12-15 μm which becomes a standard in industrial production of HgCdTe FPAs today [21-23].

Within last two years, there has been a strong effort to develop FPAs with pixel size approaching the diffraction limit initiated by the program of the US Advanced Wide FOV Architecture for Image Reconstruction and Exploitation (AWARE)-Lambda Scale detector proposed by Dr. Nibir Dhar of the Defense Advanced Research Projects Agency (DARPA). In that program, detector pixel size was proposed to be reduced to 5μm for both, LWIR and MWIR HgCdTe FPAs [24]. This task presented serious challenges associated with: detector processing technologies, lower optical efficiency, larger cross-talk between pixels, interconnect density, and high charge capacity. In spite of these challenges Teledyne has demonstrated recently small 40x40 test HgCdTe LWIR FPAs with pixel size of 5μm [25]. Interestingly, these devices demonstrate significantly better performance than that established by "Rule 07" trend line [26]. Authors attribute this effect to lower doping and thinner active layer in optimized architecture resulting in reduction of Auger-limited diffusion dark currents. More recently, full size, 5μm pitch 1280x720 pixel MWIR and LWIR FPAs have been reported by researchers from DRS Technologies [27]. Such a small pitch was achieved by downscaling DRS' HDVIP® architecture with interconnects to silicon readout chip obtained by dry etched and metalized vertical vias. The collection efficiency, cross-talk, and operability are shown to be similar to larger pitch HDVIP FPAs reported by these authors previously [21]. They also observed lower dark current levels (factor of 2-4) in comparison to those measured in larger, 12μm and 15μm pitch FPAs [21]. This is in agreement with already mentioned observations reported by Teledyne researchers [21]. These new developments allow for arrays with density of 4 million pixels/cm^2 and production of lower cost FPAs from HDTV resolution up to many millions of pixels.

Investigating the FPA scaling-down even further, Kinch considered noise equivalent temperature difference (NEDT) in diffraction limited optical system [28]. Assuming ideal detector, with zero dark current and optics defined by the relation $F\lambda/d>2$ between F number, pixel size d, and operating wavelength λ, required for high quality image, he found that optics limited NEDT attains its minimum of ~30mK with ~2.5μm pitch and F~0.5 optics for both MWIR and LWIR bands. After including typical dark current levels for different available today detector architectures (n-on-p, p-on-n, barrier detector based on III-V materials, and PIN diode) he concluded that all these technologies can deliver background limited performance (BLIP) if operating below 150K and 120K for MWIR and LWIR devices, respectively. At 300K however, only HgCdTe PIN diode technology allows for optics limited performance with absorber doping below $2x10^{14}cm^{-3}$ level needed for its depletion and hence elimination of Auger dark current component. The dark current in other considered technologies needs to be significantly reduced to allow optics limited performance at higher operating temperatures (ultimately room temperature). Theoretically, type-II InAs/InAsSb superlattice nBn barrier structure, suppressing Auger diffusion dark current, could meet requirements if only SRH lifetime > 200µs was achievable. Unfortunately, at present it is < 1µs. On the other hand, HgCdTe material meets this lifetime requirement and therefore nBn detector architecture has been intensively studied in this material in recent years. This subject is discussed in the next section.

Reduction of the pixel size and pitch during recent years is impressive. The size comparison of 1280x720 FPA with 12µm, 5µm, and projected 2.5µm pitch is presented in Fig. 3. Rough estimate shows that around 50 FPA units with 1280x720 resolution and 12µm pitch can be processed on 4" wafer, assuming that only 75% of the wafer area is usable. In case of the same FPA with 5µm pitch this number raises to 250 and for 2.5µm pitch increases to a staggering 1000 unites per 4" wafer. It is quite possible that with such a high number of FPAs per wafer and increasing detector operation temperature, the need for ever larger wafers, matching silicon readout CTE, may disappear in the near future in favor of already reported 4" (possible larger in near future) CdZnTe wafers.

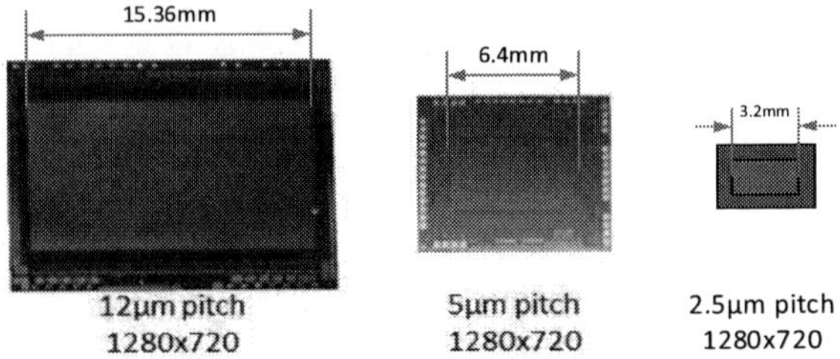

Fig. 3 Size comparison of 12µm (left), 5µm FPA (center), and projected 2.5µm (right) pitch FPAs (after Ref. 27). The FPA chip is placed on slightly larger silicon readout with bonding pads visible at the edges.

BARRIER DETECTOR ARCHITECTURE

A new IR detector architecture based on a unipolar nBn structure has been recently proposed to suppress dark current and thus enhance the operating temperature. Figure 4 shows the schematic energy band diagram of an nBn detector [29,30] which comprises an n-type narrow band gap absorber region coupled to a thin wide bandgap barrier layer, followed by a narrow bandgap contact region. The critical point to the application of this nBn device architecture is zero valence band offset (no barrier for minority holes) while maintaining a large conduction band offset (electron barrier). Such a barrier arrangement allows photogenerated minority carrier holes to flow to the contact unimpeded, even at very low bias, while the majority carrier dark current, re-injected photocurrent, and surface current are blocked by the large energy barrier in the conduction band. Because there is no high field depletion region present, nBn detector can effectively suppress dark current associated with SRH processes as well as tunneling current and noise without impeding the photocurrent (signal). In addition, the large energy barrier in the conduction band can also suppress surface leakage current, which mitigates the need for high quality surface passivation. Due to its barrier design nBn detector behaves like a photoconductor except for the presence of the barrier. Therefore, its fabrication process is simpler than that of photodiodes resulting in higher pixel yield and reduced cost. Theoretically, nBn detectors, like photoconductors, should be insensitive to point defects, which should also result in improved uniformity and higher pixel yield in comparison to photovoltaic detectors.

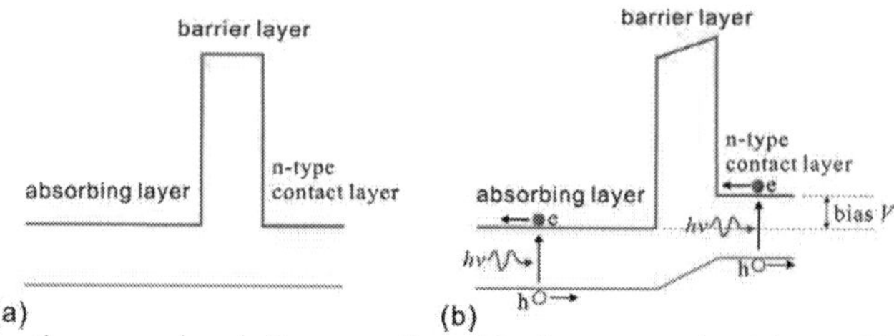

Fig. 4. Schematic energy band diagram of an nBn detector under (a) zero bias, and (b) a low reverse bias V (after Ref.29).

The nBn device architecture was first proposed by Maimon and Wickst [29], and has been widely studied for III-V material systems such as InGaAs and InAs/GaSb type−II superlattice [31-33]. Megapixel MWIR and LWIR cryogenically cooled FPAs based on type-II superlattice nBn device architecture have been demonstrated with excellent imaging quality [34-36].

As mentioned in the previous section, barrier detectors based on III-V materials so far can't meet requirements for HOT operation due to short SRH lifetime achievable today. HgCdTe SRH lifetime doesn't have this problem, but in contrast to III-V materials implementation of nBn architecture in this material presents a serious challenge due to the difficulty in realizing an ideal nBn band diagram with zero valence band offset (no hole barrier), combined with a large conduction band offset (effective electron barrier). Figure 5 shows the energy band diagram of an nBn HgCdTe detector with high x value $Hg_{1-x}Cd_xTe$ barrier layer. The existence of a valence band offset in HgCdTe−based nBn detectors seriously limits the device performance [37-40]. At low bias, the valence band barrier inhibits minority holes flowing from absorber to contact cap layer. Depending on the wavelength of operation, a relatively high bias, usually higher than bandgap energy, is required to collect all of the photogenerated carriers. This leads to the formation of a depletion layer and band-to-band and trap-assisted tunneling due to high electric field. MWIR HgCdTe devices with such an nBn structure (with valence band offset) and 5.7µm

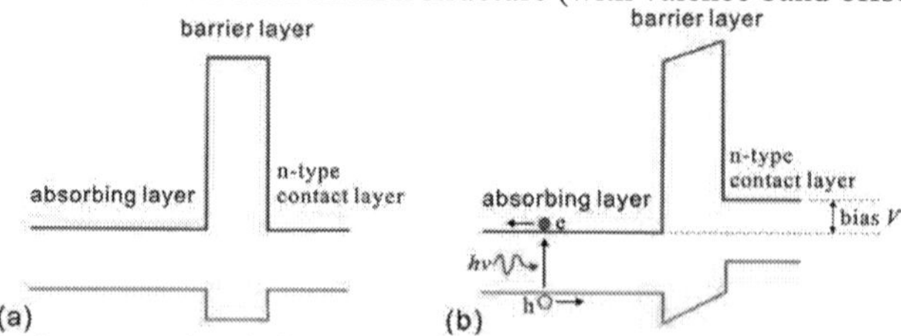

Fig. 5. Schematic energy band diagram of an nBn HgCdTe detector with high x value $Hg_{1-x}Cd_xTe$ alloy as the barrier layer: (a) zero bias and (b) under low reverse bias V.

cut−off wavelength have been reported [37]. Figures 6 (a) and (b) present the measured current density as a function of bias and temperature, respectively. Within the temperature range between 180K and 250K, the dark current corresponds to the calculated model for a diffusion limited device. Below 180K, the experimental dark current saturates due to carrier generation via surface trap states along the sidewalls in the narrow bandgap absorber induced by dry mesa etching. A "turn-on" voltage of – 0.5 V to – 1.0 V, indicates detrimental effect of valence band offset.

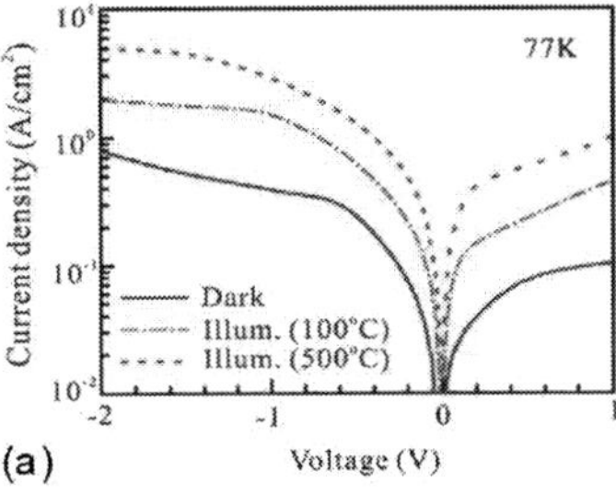
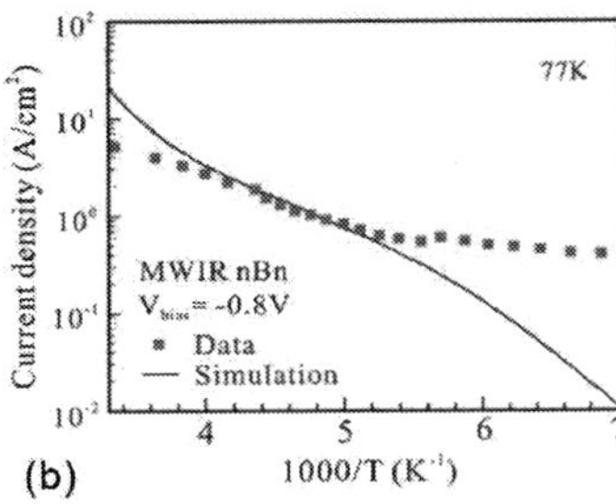

Fig. 6. (a) I-V characteristics, and (b) current density as a function of temperature for MWIR nBn HgCdTe detector with $Hg_{0.36}Cd_{0.64}Te$ barrier layer (after Ref.38).

Among possible approaches to reduce or even eliminate valence band offset in HgCdTe based barrier detectors are shaping the composition and doping profile within a barrier, or use of a superlattice barrier. Both these approaches have been studied theoretically. Considering barrier doping/composition shaping approach, Kopytko et al. [41] reported that optimisation of barrier doping and composition leads to reduction of the valence band offset to 60meV for MWIR, and 50meV for LWIR structures, respectively with conduction band barrier sufficiently high to effectively block transport of electrons in both MWIR and LWIR structures as shown in Figure 7. Akhavan et al.

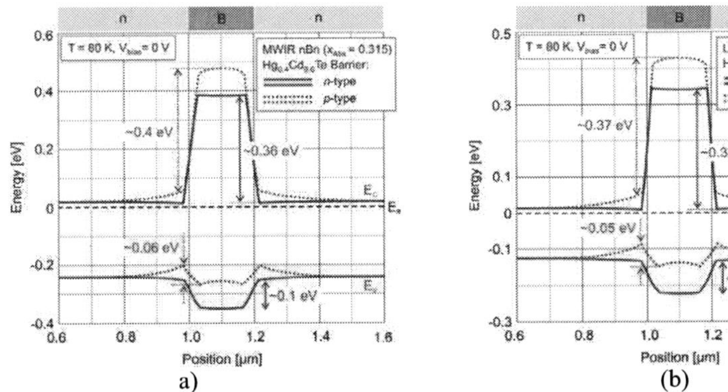

Fig. 7. Comparison of equilibrium band diagrams for (a) MWIR and (b) LWIR HgCdTe nBn detector with n-type ($N_D = 1\times10^{15}$ cm^{-3}) and p-type ($N_A = 3\times10^{15}$ cm^{-3}) doped barrier layers (after Ref. 41).

also studied the doping/composition shaping approach theoretically and concluded that efficient elimination of the valence band discontinuity is possible allowing the device to operate with $|V_{bias}| < 50$ mV [42,43]. This result renders all tunneling-related dark current components insignificant and allows the detector to achieve the maximum possible diffusion current limited performance. In more recent article Akhavan et al. considered theoretically operation of such a nBn structure at higher temperatures [44]. The results obtained indicate that the composition, doping, and thickness of the barrier layer in MWIR HgCdTe nBn detectors can be optimized to yield performance levels comparable with ideal HgCdTe p–n photodiodes. They also show that introduction of an additional barrier at the back contact layer (nBnn[+]) leads to substantial suppression of the Auger generation–recombination (GR) mechanism resulting in an order-of-magnitude reduction in the dark current compared with conventional nBn or p–n junction-based detectors, enabling background-limited operation above 200 K. Such a HgCdTe nBn detectors with 3.5µm cut-off wavelength have been reported by Kopytko et al. [45]. Figure 8 shows their spectral responsivity characteristics at high temperatures demonstrating no bias dependence indicating successful elimination of valence band offset. Similar LWIR devices have not been reported yet.

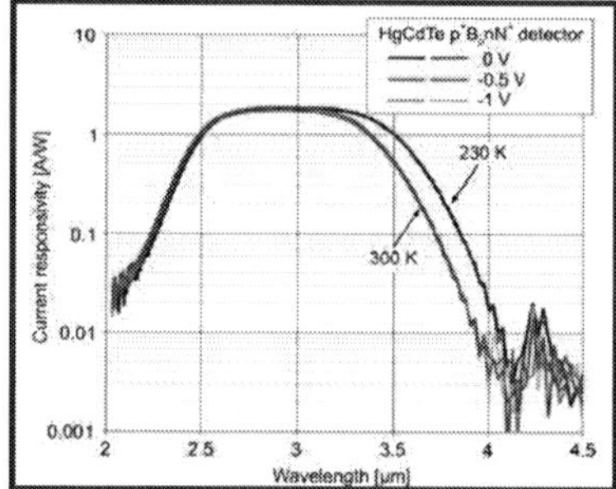

Fig. 8. Spectral responsivity characteristics of 3.5µm cut-off HgCdTe nBn detectors operating at high temperatures (after Ref. 45).

The second, SL based, approach for removing valence band offset in HgCdTe nBn structures provides more flexibility. Figure 9 illustrates the energy band diagrams calculated for nBn structures with different superlattice barrier designs.

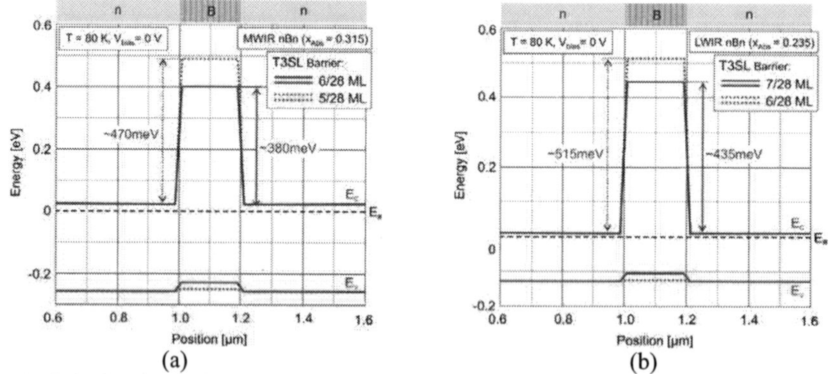

(a) (b)

Fig. 9. Calculated zero bias energy band diagrams for (a) MWIR and (b) LWIR HgCdTe nBn detector with HgTe/Hg$_{0.05}$Cd$_{0.95}$Te type-3 SL barrier layers (after Ref. 41).

Analysing their calculations authors conclude that nBn detector, with a barrier formed using a 5ML thick HgTe (well) and 28ML $Hg_{0.05}Cd_{0.95}Te$ (barrier) represents a close to optimal design that meets the requirements for a HgCdTe-based high-performance nBn detector technology for MWIR spectral range. Corresponding conditions for LWIR devices are 6ML HgTe/28ML $Hg_{0.05}Cd_{0.95}Te$. The fabrication of SL based nBn detectors have not been reported yet and it will be interesting to see if real devices will confirm theoretical predictions.

HIPERSPECTRAL DETECTORS

Apart from multi-band detector technologies consisting of a stack of two detector layers separated by a wide bandgap common electrode and hence registering simultaneously IR from two different spectral bands [46], adaptive MEMS filter–based IR FPA technology provides another approach to achieve multi-band detection [47,48]. The technology is built on MEMS-based electrostatically actuated Fabry–Perot adaptive filters integrated with IR FPA. Figure 10 shows the general concept and an example of simulated transmission spectrum [49,50]. In recent decade, significant effort has been

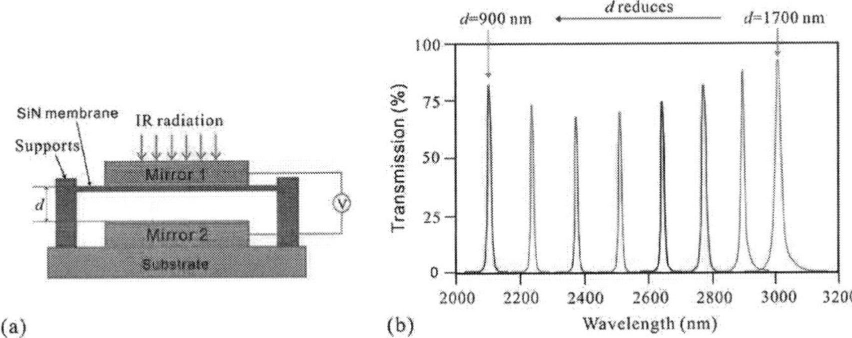

Fig. 10. (a) schematic concept and (b) simulated transmission spectra with different spacing d for a SWIR tuneable MEMS optical filter (after Refs. 49,50).

devoted to design and fabrication of such devices. Figure 11 shows experimental IR spectra reported by University of Western Australia (UWA) in SWIR, MWIR and LWIR spectral bands [49-51]. The realization of adaptive concepts offers the potential approach to achieving real- time tunable multiband detection which, under some circumstances, is more beneficial than broad-band multi-color detection. Despite the great advances that have been made, the technology is still very challenging. The main issues include: *(i)* parallel or flat actuating mirror in the whole deflection range, *(ii)* larger tuning range to enable detection of broader spectrum, *(iii)* robust actuating membrane structure, and *(iv)* capability of operating at cryogenically-cooled temperatures. Obviously, the current research effort in this field is relatively small compared with that of other multi-band detection technologies, and more effort is required in order to address these significant technological challenges. However, it is important to note that this new technology is currently being extended to large area adaptive filter designs (multi-mm × multi-mm). This development when combined with new small in size high pixel density diffraction limited FPAs discussed earlier in this article will provide the means for fabrication of adaptive, hyperspectral high pixel density FPAs.

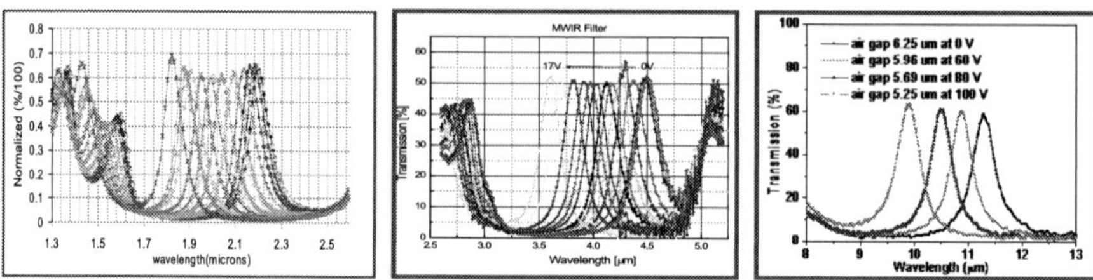

Fig. 11. Experimental optical spectra of MEMS based IR tunable filters fabricated at UWA (after Refs. 49-51).

CONCUSIONS

In spite of many challenges and competition over the years, HgCdTe technology withstood the test of time and today is approaching mature age. It is undoubtedly the material and technology of choice for high performance IR detectors and FPAs. Long lasting problem with large size high quality substrates fades away with the increasing quality of alternative substrates, but particularly due to scaling down of the FPA pixel size and pitch down to diffraction limit and development of large lattice matched CdZnTe substrates. The size of fully operational HDTV IR HgCdTe FPAs has been reduced from 32mm x 18mm for 25μm pitch technology used in commercial production today down to reported recently 6.4mm x 3.6mm with 5μm pitch, and it is expected to be scaled down to 3.2mm x 1.8mm when the pitch of the FPA is further reduced to optimal diffraction limited size of 2.5μm. With recently reported CdZnTe epi-ready 4" substrates these two developments allow for processing of up to 250 5μm pitch FPAs from one 4" wafer, and staggering ~1000 FPAs when 2.5μm technology is developed. Usually observed non-uniformity of pixel's cut off wavelength due to growth related slow fluctuations of HgCdTe composition (and hence energy gap) across relatively large area of present FPAs is expected to improve with introduction of new small area but high density devices.

The major remaining challenge is improvement of HgCdTe material and device architecture enabling lifting of the operation temperature above 200K where much cost effective thermos-electric coolers can be used. Due to superior transport and optical properties of HgCdTe material, optimized unipolar barrier architectures based on this material seem to be a promising concept. Room operating single devices with relatively short cut-off wavelength (3.5μm) have been already reported, but it remains to be seen if the theoretically predicted performance of MWIR and LWIR devices will be confirmed experimentally.

ACKNOWLEDGEMENTS

This work was supported by the Australian Research Council (FT130101708, DP140101766, LP120100097, LP120200764, DP120104835, DP150104839 and

DP0881579,), and Go8-DAAD collaboration program (2014-2015), and a Research Development Award from the Faculty of Engineering, Computing and Mathematics at The University of Western Australia.

REFERENCES

1. M. Reddy, J.M. Peterson, T. Vang, J.A. Franklin, M.F. Vilela, K. Olsson, E.A. Patien, W.A. Radford, J.W. Bangs, L. Melkonian, E.P.G. Smith, D.D. Lofgreen and S.M. Johnson, *J. Electron. Mater.* , **40**, 1706 (2011).
2. M. Carmody, A. Yulius, D. Edwall, D. Lee, E. Piquette, R. Jacobs, D. Benson, A. Stoltz, J. Markunas, A. Almeida and J. Arias, *J. Electron. Mater.* **41**, 2719 (2012).
3. J. Wenisch, D. Eich, H. Lutz, T. Schallenberg, R. Wollrab and J. Ziegler, *J. Electron. Mater.* **41**, 2828 (2012).
4. L. He, L. Chen, Y. Wu, X.L. Fu, Y.Z. Wang, J. Wu, M.F. Yu, J.R. Yang, R.J. Ding, X.N. Hu, Y.J. Li and Q.Y. Zhang, *Journal of Crystal Growth*, **301-302**, 268 (2007).
5. J.P. Zanatta, G. Badano, P. Ballet, C. Largeron, J. Baylet, O. Gravrand, J. Rothman, P. Castelein, J.P. Chamonal, A. Million, G. Destefanis, S. Mibord, E. Brochier and P. Costa, *J. Electron. Mater.* **35**, 1231 (2006).
6. W. Lei, R.J. Gu, J. Antoszewski, J. Dell and L. Faraone, *J. Electron. Mater.* **43**, 2788 (2014).
7. G. Brill, S. Farrell, Y.P. Chen, P.S. Wijewarnasuriya, M.V. Rao, J.D. Benson and N. Dhar, *J. Electron. Mater.* **39**, 967 (2010).
8. J. Garland and R. Sporken, Chapter 4 in book *Mercury Cadmium Telluride: Growth, Properties and Applications*, (Edited by P.C.a.J.W. Garland, *John Wiley & Sons Ltd West Susses. 2011*), p.75-90.
9. S. Farrell, G. Brill, Y. Chen, P.S. Wijewarnasuriya, M.V. Rao, N. Dhar and K. Harris, *J. Electron. Mater.* **39**, 43 (2010).
10. S. Farrell, M.V. RAO, G. BRILL, Y. CHEN, P. Wijewarnasuriya, N. Dhar, D. Benson and K. Harris, *J. Electron. Mater.* **40**, 1727 (2011).
11. S. H. Shin, J. M. Arias, D. D. Edwall, M. Zandian, J. G. Pasko and R.E. DeWames, *J. Vac. Sci. Technol. B*, **10**, 1492 (1992).
12. A.J. Stoltz, J.D. Benson, M. Carmody, S. Farrell, P.S. Wijewarnasuriya, G. Brill, R. Jacobs and Y. Chen, *J. Electron. Mater.* **40**, 1785 (2011).
13. A.J. Stoltz, J.D. Benson, R. Jacobs, P. Smith, L. A. Almeida, M. Carmody, S. Farrell, P. S. Wijewarnasuriya, G. Brill and Y. Chen, *J. Electron. Mater.* **41**, 2949 (2012).
14. R. Jacobs, A.J. Stoltz, J.D. Benson, P. Smith, C.M. Lennon, L. A. Almeida, S. Farrell, P. S. Wijewarnasuriya, G. Brill, Y. Chen, M. Salmon and J. Zu, *J. Electron. Mater.* **42**, 3149 (2013).
15. J.D. Benson, L.O. Bubulac, P.J. Smith, R.N. Jacobs, J.K. Markunas, M. Jaime-Vasquez, L.A. Almeida, A. Stoltz, J.M. Arias, G. Brill, Y. Chen, P.S. Wijewarnasuriya, S. Farrell and U. Lee, *J. Electron. Mater.* **41**, 2971 (2012).
16. R.N. Jacobs, C. Nozaki, L.A. Almeida, M. Jaime-Vasquez, C. Lennon, J.K. Markunas, D. Benson, P. Smith, W.F. Zhao, D.J. Smith, C. Billman, J. Arias and J. Pellegrino, **41**, *J. Electron. Mater.* 2707 (2012).

17. L. He, X. Fu, Q. Wei, W. Wang, L. Chen, Y. Wu, X. Hu, J. Yang, Q. Zhang, R. Ding, X. Chen and W. Lu, *J. Electron. Mater.* **37**, 1189 (2008).
18. W. Lei, R.J. Gu, J. Antoszewski, J. Dell, G. Neusser, M. Sieger, B. Mizaikoff and L. Faraone, *J. Electron.Mater.* online (June, 2015)
19. J. Mackenzie, F.J. Kumar, H. Chen. J. Electron. Mater. **42**, 3129 (2013).
20. D. Brellier, E. Gout, G. Gaude, D. Pelenc, P. Ballet, T. Miguet, M.C. Manzato, J. Electron. Mater. **43**, 2901 (2014).
21. R.L. Strong, M.A. Kinch and J. Armstrong, *J. Electron. Mater.* **42**, 3103 (2013).
22. J. Wenisch, H. Bitterlich, M. Bruder, P. Fries, R. Wollrab, J. Wendler, R. Breiter and J. Ziegler, *J. Electron. Mater.* **42**, 3186 (2013).
23. O. Gravrand and G. Destefanis, *Infrared Phys. Technol.*, **59**, 163 (2013).
24. N. Dhar, R. Dat, A. Sood, InTech, open access article, http://cdn.intechopen.com/pdfs-wm/38815.pdf
25. W.E. Tennant, D.J. Gulbransen, A. Roll, M. Carmody, D. Edwall, A. Julius, P. Dreiske, A. Chen, W. Mclevige, S. Freeman, D. Lee, D.E. Cooper and E. Piquette, *J. Electron. Mater.* **43**, 3041 (2014).
26. W.E. Tennant, J. Prog. Quantum Electron. **36**, 273 (2012).
27. J. M. Armstrong ; M. R. Skokan ; M. A. Kinch ; J. D. Luttmer, Proc. SPIE 9070, Infrared Technology and Applications XL, 907033 (June 24, 2014).
28. M. Kinch, J. Electron. Mater. Online, (March 2015).
29. S. Maimon and G.W. Wicks, *Appl. Phys. Lett.*, **89**, 151109 (2006).
30. P. Martyniuk, M. Kopytko and A. Rogalski, *Opto-Electronics Review*, **22**, 127 (2014).
31. A. Khoshakhlagh, S. Myers, H. Kim, E. Plis, N. Gautam, S.J. Lee, S. K. Noh, L.R. Dawson and S. Krishna, *IEEE Journal of Quantum Electronics*, **46**, 959 (2010).
32. H. S. Kim, O. O. Cellek, Z.Y. Lin, Z.Y. He, X.H. Zhao, S. Liu, H. Li and Y.H. Zhang, *Appl. Phys. Lett.*, **101**, 161114 (2012).
33. A. Khoshakhlagh , J.B. Rodriguez, E. Plis, G.D. Bishop, Y.D. Sharma, H.S. Kim, L.R. Dawson and S. Krishna, *Appl. Phys. Lett.* , **91**, 263504 (2007).
34. C.J. Hill, A. Soibel, S.A. Keo, J.M. Mumolo, D.Z. Ting, S.D. Gunapala, D.R. Rhiger, R.E. Kvaas and S.F. Harris, *Proc. SPIE*, **7298**, 7294 (2009).
35. S.D. Gunapala, D.Z. Ting, C.J. Hill, J. Nguyen, A. Soibel, S.B. Rafol, S.A. Keo, J.M. Mumolo, M.C. Lee, J.K. Liu and B. Yang, *Phot. Tech. Lett.* **22**, 1856 (2010).
36. P. Manurkar, S. Ramezani−Darvish, B.M. Nguyen, M. Razeghi and J. Hubbs, *Appl. Phys. Lett.*, **97**, 193505 (2010).
37. A.M. Itsuno, J.D. Phillips and S. Velicu, **40**, 1624 (2011).
38. A. M. Itsuno, J.D. Phillips and S. Velicu, *Appl. Phys. Lett.*, **100**, 161102 (2012).
39. S. Velicu, J. Zhao, M. Morley, A. M. Itsuno and J.D. Phillips, *SPIE Proc.* **8268**, 82682X (2012).
40. P. Maryniuk and A. Rogalski, *Solid−State Electronics*, **80**, 96 (2013).
41. M. Kopytko, J. Wrobel, K. Kozwikowski, A. Rogalski, J. Antoszewski, N.D. Akhavan, G.A. Umana-Membreno, L. Faraone, C.R. Becker, J. Electron. Mater. **44**, 158 (2015)
42. N. D, Akhavan, G. Jolley, G.A. Umana-Membreno, J. Antoszewski, L. Faraone, IEEE Transactions on Electron Devices, Electron Devices, **61**, 3691 (2014)
43. N. D, Akhavan, G. Jolley, G.A. Umana-Membreno, J. Antoszewski, L. Faraone, IEEE Transactions on Electron Devices, Electron Devices, **62**, 722 (2015)

44. N. D, Akhavan, G. Jolley, G.A. Umana-Membreno, J. Antoszewski, L. Faraone, J. Electron. Mater. Online (April 2015)
45. M. Kopytko, A. Keblowski, W. Gawron, A. Kowalewski, A. Rogalski, IEEE Trans. Electron. Dev. **61**, 3691 (2014)
46. A. Rogalski, J. Antoszewski and L. Faraone, *J. Appl. Phys.*, **105**, 091101 (2009).
47. J. Carrano, J. Brown, P. Perconti and K. Barnard, *Tuning in to detection*, in *SPIE's OEmagazine*. p. p.20-22.
48. A. Rogalski, *Acta Physica Polonica A*, **116**, 389 (2009).
49. L. Faraone, *Proc. of SPIE*, **5957**, 59570F (2005).
50. C. A. Musca, J. Antoszewski, A. J. Keating, K. J. Winchester, K. K. M. B. D. Silva, T. Nguyen, J. M. Dell and L. Faraone, *Proceeding of 2007 IEEE/LEOS International Conference on Optical MEMS and Nanophotonics (Aug. 12-16, 2007)*, Hualien, p.137
51. H.F. Mao, K.K.M.B. D. Silva, M. Martyniuk, J. Antoszewski, J. Bumgarner, B.D. Nener, J.M. Dell and L. Faraone, (unpublished data).

ECS Transactions, 69 (14) 77-82 (2015)
10.1149/06914.0077ecst ©The Electrochemical Society

Flexible Graphene Electrode-based Organic Photovoltaics with Record-High Efficiency

H. Park[a,b,c], S. Chang[b], X. Zhou[b], J. Kong[a], T. Palacios[a], and S. Gradečak[b]

[a] Department of Electrical Engineering and Computer Science, Massachusetts Institute of Technology, Cambridge, Massachusetts 02139, USA
[b] Department of Materials Science and Engineering, Massachusetts Institute of Technology, Cambridge, Massachusetts 02139, USA
[c] School of Energy and Chemical Engineering, Ulsan National Institute of Science and Technology, Ulsan, 689-798, South Korea

Advancements in the field of flexible high-efficiency solar cells and other optoelectronic devices will strongly depend on the development of electrode materials with good conductivity and flexibility. To address chemical and mechanical instability of currently used indium tin oxide (ITO), graphene has been suggested as a promising flexible transparent electrode, but challenges remain in achieving high efficiency of grahene-based polymer solar cells (PSCs) compared to their ITO-based counterparts. Here we demonstrate graphene anode- and cathode-based flexible PSCs with record-high power conversion efficiencies of 6.1% and 7.1%, respectively. The high efficiencies were achieved via thermal treatment of MoO_3 electron blocking layer and direct deposition of ZnO electron transporting layer on graphene. We also demonstrate graphene-based flexible PSCs on polyethylene naphthalate substrates and show the device stability under different bending conditions. Our work paves a way to fully graphene electrode-based flexible solar cells using a simple and reproducible process.

Introduction

Flexible organic and hybrid organic-inorganic solar cells require both the flexible photoactive media[1,2] and flexible electrodes with good conductivity and transparency. So far, indium tin oxide (ITO) has been an electrode material of choice for studies focusing on optimizing the morphology and chemistry of the photoactive media[3] in polymer solar cells (PSCs). For flexible applications, graphene has been proposed as a promising replacement for ITO due to its mechanical and chemical robustness, excellent electrical and optical properties, and potentially low-cost processing[4,5]. Recent studies have demonstrated dramatic improvements in the efficiency of PSCs[6,7] that ascertain the bright future towards PCE≥10%, a threshold considered for industrial applications[2]. In PSCs, and more generally organic solar cells (OSCs), one of the electrodes typically consists of a transparent conductor, among which ITO is most widely used due to its good optical transparency and electrical conductivity. However, due to its non-uniform absorption, chemical and mechanical instability, as well as high cost of indium[8], several alternative materials have been proposed, including carbon nanotube[9,10] or metallic nanowires networks[11]. Furthermore, even at small mechanical stresses, microcracks are produced in ITO resulting in increased film resistance and decreased device performance[12], thus limiting its applications for flexible OSCs. Owing to the unique optoelectronic properties of graphene[5], several works on graphene-based OSCs have been reported, demonstrating the feasibility of graphene in transparent electrode applications[3,12-15]. While the initial demonstrations have been promising, performance of graphene-based devices still falls short (< 3%) of recent advances accomplished in ITO-based devices (8 – 9%)[6,7]. Therefore, to demonstrate graphene as an emerging alternative to ITO, it is inevitable to improve the currently low-efficiency of graphene-based solar cells. Moreover, to test feasibility of graphene for flexible applications, high performance and stability of such devices must be investigated. In this work, we demonstrate high-efficiency graphene anode- and cathode-based PSCs with PCEs comparable to their ITO-counterparts. We also demonstrate graphene-based flexible PSCs on polyethylene naphthalate (PEN) substrates and show the device stability under different bending conditions.

Graphene Anode-based Polymer Solar Cells

The overall PSC device structure used in this work and the corresponding band diagram are shown in Fig. 1a-b. To fabricate efficient graphene-based OSCs, we used photoactive media composed of a blend of low bandgap semiconducting polymer donor thieno[3,4-b]thiophene/benzodithiophene (PTB7) and acceptor [6,6]-phenyl C_{71}-butyric acid methyl ester ($PC_{71}BM$) prepared using mixed solvents of chlorobenzene:1,8-diiodoctane (CB:DIO, 97:3 vol%)[6]. The hole injection layer poly(3,4-ethylenedioxythiophene):poly(styrenesulfonate) (PEDOT:PSS) was deposited on the transparent graphene electrode. To ensure uniform coverage over the graphene surface, we used modified PEDOT:PSS with isopropyl alcohol (IPA) at 3:1 (v/v) ratio. Prior to the active device layer deposition, graphene/PEDOT:PSS must be covered by an additional electron blocking layer (MoO_3). This process is critical in graphene-based bulk heterojunction PSCs since the additional MoO_3 layer helps prevent charge recombination occurring at the graphene/PEDOT:PSS and the polymer blend interface[16]: graphene/PEDOT:PSS only devices (i.e., without the MoO_3 layer) lead to shorting in the device performance. We also note that although ITO does not require additional MoO_3 layer, ITO control devices were fabricated with PEDOT:PSS/MoO_3 for direct comparison with graphene-based devices.

We discovered that one critical aspect of the device fabrication is the thermal annealing of MoO_3 layer before spin-coating $PTB7:PC_{71}BM$. When directly applying $PTB7:PC_{71}BM$ to PEDOT:PSS/MoO_3 on ITO or graphene without thermal annealing, considerable degradation in device performance was observed, which did not occur on PEDOT:PSS only device. Fig.1c presents the current density–voltage (J–V) characteristics of PSCs fabricated on ITO substrates, in which addition of MoO_3 layer (non-annealed) results in detrimental effects on the short-circuit current density (J_{SC}, 2.6 mA/cm^2), open-circuit voltage (V_{OC}, 0.57 V), and fill factor (FF, 31.3%). However, significant improvements in J_{SC} (16.1 mA/cm^2), V_{OC} (0.68 V), FF (60.7%), and the resulting increase in PCE were observed after the MoO_3 electron blocking layer thermal annealing.

Graphene-based PSCs were fabricated under the same experimental conditions and the resulting J–V characteristics are compared with that of ITO reference device in Fig. 1d. The graphene electrode was made by stacking three monolayers of graphene film prepared by low pressure chemical vapor deposition, with a typical sheet resistance of

~300 Ω/sq and transmittance of ~92% at λ = 550 nm[13]. With incorporation of appropriate PEDOT:PSS and thermally-treated MoO_3, we observed record-high efficiency from graphene (PCE=6.1%) approaching to ITO reference device (PCE=6.7%).

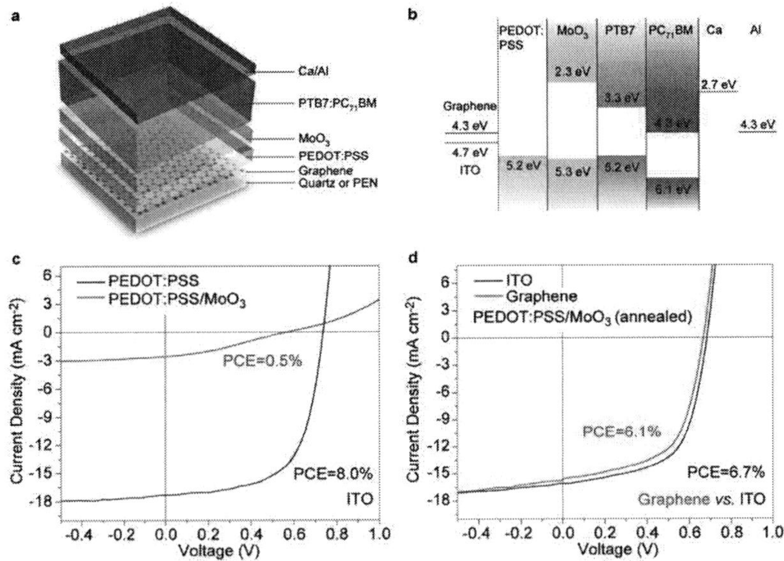

Figure 1. Device structure, energy levels, and performance of graphene- and ITO-based PSCs on quartz. (**a**) Schematic of a graphene anode-based PSC, in which additional MoO_3 electron blocking layer is inserted between the conventional PEDOT:PSS hole injection layer and the PTB7:PC_{71}BM polymer blend. (**b**) Corresponding flat-band energy level diagram. (**c**) AM1.5G $J–V$ characteristics of ITO-based PSCs: ITO/PEDOT:PSS(40nm)/(MoO_3(20nm)/PTB7:PC_{71}BM(200nm)/Ca(25nm)/Al(80nm). (**d**) AM1.5G $J–V$ characteristics of graphene and ITO anode-based PSCs: anode (graphene or ITO)/PEDOT:PSS/MoO_3(annealed at 150°C for 10 min)/PTB7:PC_{71}BM/Ca/Al).

<u>Graphene Cathode-based Polymer Solar Cells</u>

In addition to the graphene anode-based devices discussed above, we also investigated inverted cathode-based PSC configuration that can provide improved device stability by avoiding easily oxidized low work function metal electrodes such as Al or $Ca^{8,17}$. In this structure, n-type semiconducting metal oxides such as ZnO or TiO_x can be utilized as an effective electron transporting path from the photoactive polymer layer to the cathode. After achieving uniform coverage of the electron transporting ZnO layer directly on graphene surface by a simple spin-coating method, graphene cathode-based inverted PSCs and their ITO-based counterparts were demonstrated with the following device structure: graphene (or ITO)/ZnO/PTB7:PC_{71}BM/MoO_3/Ag. The device structure, corresponding band diagram, and measured $J–V$ characteristics are shown in Fig. 2. As illustrated in Fig. 2c, both graphene- and ITO-based inverted solar cells show great similarities in device performance with PCEs of 6.9% and 7.6%, respectively. We note that the inverted cathode-based PSCs fabricated on graphene electrodes without the ZnO layer exhibit negligible photoresponse, indicating the importance of electron transporting layer on the graphene cathode.

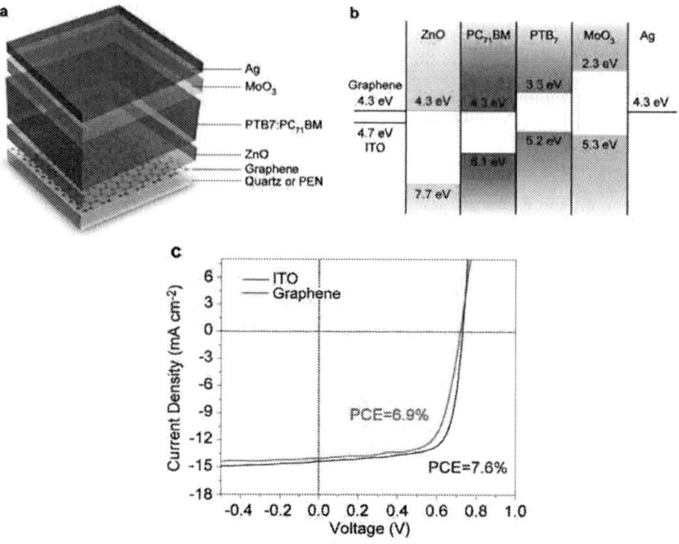

Figure 2. Graphene and ITO cathode-based PSCs: cathode (graphene or ITO)/ZnO(20nm)/PTB7:PC$_{71}$BM(200nm)/MoO$_3$(20nm)/Ag(100nm). (a) Schematic of inverted PSCs and (b) corresponding flat-band energy level diagram. (c) AM1.5G J–V characteristics of a graphene-based device, demonstrating comparable performance to that of an ITO reference cell.

<u>Graphene-based Flexible Polymer Solar Cells</u>

With growing interests in flexible electronic devices[18], as the final step we have explored the potential of our device structures to realize flexible graphene-based PSCs. We have realized both anode- and cathode-based device architectures on PEN substrates and tested their performance under bending conditions. The completed devices were tested under mechanical tensile bending conditions, which were subject to consecutive flexing cycles (at 3.5 mm radius) with pre-determined bending strain (ε) of ~4.3%. The resulting graphene PSCs on PEN substrates show excellent device performance for both anode (PCE=6.1%, Fig. 3a) and cathode (PCE=7.1%, Fig. 3b) configurations. Our graphene-based flexible PSCs are robust under mechanical deformations, which is highly desirable for low-cost productions such as roll-to-roll processing and applications that require flexibility. As shown in Fig. 3c, the device (graphene anode) did not display any significant performance changes up to 100 tensile flexing cycles. Key photovoltaic parameters at intermediate flexing cycles are shown in Fig. 3e-f.

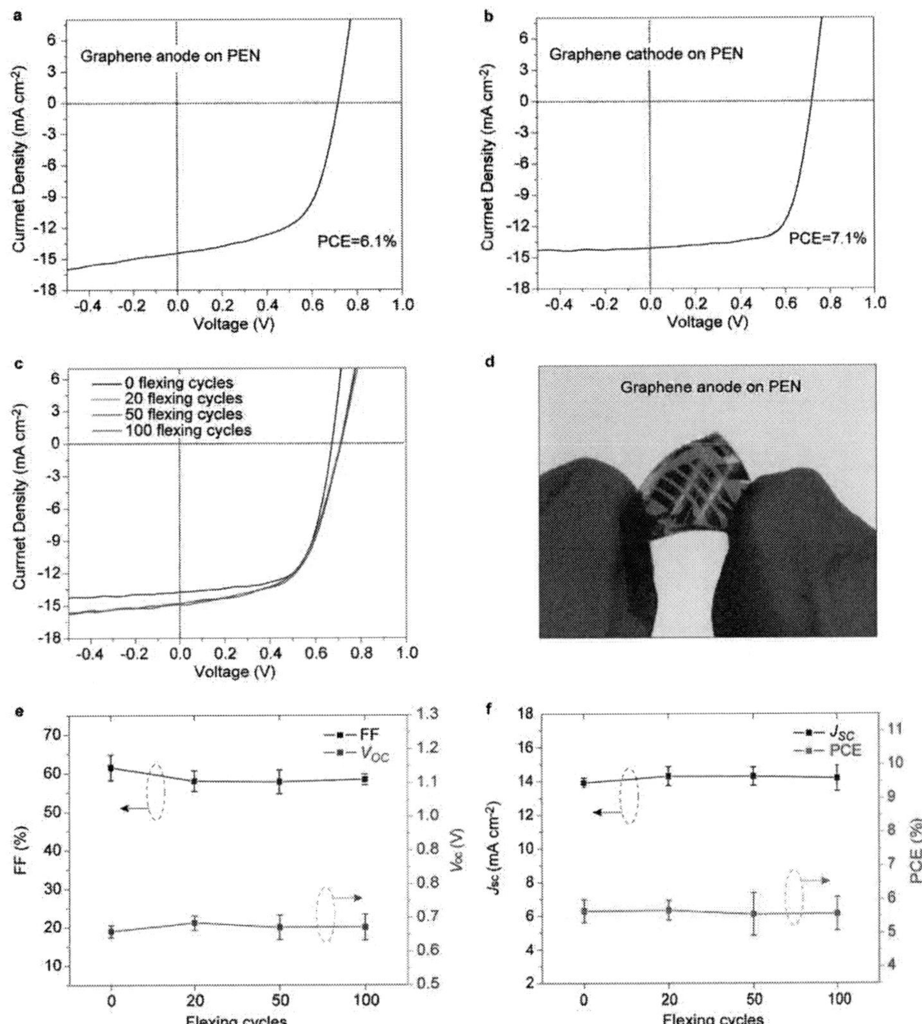

Figure 3. Graphene-based flexible PSCs on PEN. (**a**) AM1.5G *J–V* characteristics of a representative graphene anode-based device fabricated on PEN: graphene anode/PEDOT:PSS/MoO$_3$/PTB7:PC$_{71}$BM/Ca/Al. (**b**) AM1.5G *J–V* characteristics of a representative graphene cathode-based device fabricated on PEN: graphene cathode/ZnO/PTB7:PC$_{71}$BM/MoO$_3$/Ag. (**c**) AM1.5G *J–V* characteristics of the champion graphene devices before and after different flexing cycles. (**d**) A digital photograph of a flexible graphene PSC. (**e-f**) Photovoltaic performance characteristics (J_{SC}, V_{OC}, FF and PCE) of the devices displayed in (**c**). Lines serve as a guide to the eye.

Acknowledgments

This work was supported by Eni S.p.A. under the Eni-MIT Alliance Solar Frontiers Program. The authors acknowledge access to Shared Experimental Facilities provided by the MIT Center for Materials Science Engineering supported in part by MRSEC Program of National Science Foundation under award number DMR - 0213282.

References

1. G. Li, R. Zhu, and Y. Yang, *Nature Photonics*, **6**(3), 153 (2012).
2. C. Brabec, S. Gowrisanker, J. Halls, D. Laird, S. Jia, and S. Williams, *Advanced Materials*, **22**(34), 3839 (2010).
3. H. Park, S. Chang, M. Smith, S. Gradecak, and J. Kong, *Scientific Reports*, 3, 1581 (2013).
4. S. Bae, H. Kim, Y. Lee, X. Xu, J. Park, Y. Zheng, J. Balakrishnan, T. Lei, H. Ri Kim, Y. Song, Y. Kim, K. Kim, B. Ozyilmaz, J. Ahn, B. Hong, S. Iijima, *Nature Nanotechnology*, **5**(8), 574 (2010).
5. F. Bonaccorso, Z. Sun, T. Hasan, and A. Ferrari, *Nature Photonics*, **4**(9), 611 (2010).
6. Z. He, C. Zhong, S. Su, M. Xu, H. Wu, and Y. Cao, *Nature Photonics* ,**6**(9), 591 (2012).
7. Y. Zhou, C. Fuentes-Hernandez, J. Shim, J. Meyer, A. Giordano, H. Li, P. Winget, T. Papadopoulos, H. Cheun, J. Kim, M. Fenoll, A. Dindar, W. Haske, E. Najafabadi, T. Khan, H. Sojoudi, S. Barlow, S. Graham, J. Brédas, S. Marder, A. Kahn, and B. Kippelen, *Science*, **336**(6079), 327 (2012).
8. M. Jørgensen, K. Norrman, and F. Krebs, *Solar Energy Materials and Solar Cells*, **92**(7), 686 (2008).
9. M. Rowell, M. Topinka, M. McGehee, H. Prall, G. Dennler, N. Sariciftci, L. Hu, and G. Gruner, *Applied Physics Letters*, **88**(23), 233506 (2006).
10. A. Pasquier, H. Unalan, A. Kanwal, S. Miller, and M. Chhowalla, *Applied Physics Letters*, **87**(20), 203511 (2005).
11. S. De, T. Higgins, P. Lyons, E. Doherty, P. Nirmalraj, W. Blau, J. Boland, and J. Coleman, *ACS Nano*, **3**(7), 1767 (2009).
12. L. De Arco, Y. Zhang, C. Schlenker, K. Ryu, M. Thompson, and C. Zhou, *Acs Nano*, **4**(5), 2865 (2010).
13. H. Park, P. Brown, V. Buloyic, and J. Kong, *Nano Letters*, **12**(1), 133 (2012).
14. H. Park, J. Rowehl, K. Kim, V. Bulovic, and J. Kong, *Nanotechnology*, **21**(50), 505204 (2010).
15. H. Park, R. Howden, M. Barr, V. Bulović, K. Gleason, and J. Kong, *ACS Nano*, **6**(7), 6370 (2012)
16. H. Park, Y. Shi, and J. Kong, *Nanoscale*, 5, 8934 (2013).
17. S. Hau, H. Yip, N. Baek, J. Zou, K. O'Malley, and A. Jen, *Applied Physics Letters*, **92**(25), 253301 (2008).
18. D. Angmo and F. Krebs, *Journal of Applied Polymer Science*, **129**(1), 1 (2013).

ECS Transactions, 69 (14) 83-93 (2015)
10.1149/06914.0083ecst ©The Electrochemical Society

Crystalline Tetrahedral Phases $Al_{1-x}B_xPSi_3$ and $Al_{1-x}B_xAsT_3$ (T = Si, Ge) Via Reactions of $Al(BH_4)_3$ and $M(TH_3)_3$ (M = P, As)

P. Sims[a], T. Aoki[b], J. Menéndez[c], and J. Kouvetakis[a]

[a] Department of Chemistry and Biochemistry, Arizona State University, Tempe, Arizona 85287, USA
[b] LeRoy Eyring Center for Solid State Science, Arizona State University, Tempe, Arizona 85287, USA
[c] Department of Physics, Arizona State University, Tempe, Arizona 85287, USA

Crystalline $(III\text{-}V)_{1-y}(IV_2)_y$ hybrid semiconductors $Al_{1-x}B_xPSi_3$ and $Al_{1-x}B_xAsSi_3$ have been grown on Si(001) substrates *via* reactions of $Al(BH_4)_3$ and $M(SiH_3)_3$ (M = P, As). These films are characterized for structure, composition, and phase morphology The materials form by interlinking $III\text{-}V\text{-}IV_3$ tetrahedral units that construct a cubic lattice with average diamond-like symmetry containing isolated III-V pairs. This synthetic approach was extended to alloy systems in which the III-V pairs are imbedded in a Ge-matrix using $As(GeH_3)_3$. These materials are grown on Ge/Si(001) platforms and a comparative study of two growth methods is presented. One describes a low pressure CVD route *via* reactions of $Al(BH_4)_3$ and $As(GeH_3)_3$, and the other is a gas-source MBE technique combining Al atoms from an effusion cell with the germyl-arsene precursor. Both sets of films are studied using aberration-corrected STEM. Globally the results show that $Al(BH_4)_3$ acts as a carbon-free source of Al compatible with CVD protocols.

Introduction

Semiconducting alloys with the general formula $(III\text{-}V)_{1-y}(IV_2)_y$ have been garnering interest to study the unique physics and myriad possible applications these materials offer. In particular within this larger class of materials those comprised of end members with similar lattice parameters are of great interest because they allow for tunable optoelectronic properties at a fixed lattice constant. Two chronicled systems are BNC_2 and $(GaAs)_{1-y}(Ge_2)_y$. These alloys are candidate superhard and Ge lattice-matched 1.0 eV photovoltaic materials, respectively. However the implementation of these systems to practical applications are mired by fundamental issues (1,2). These issues include phase segregation, auto-doping, and compositional inhomogeneity, and until recently the lack of bonding control has impeded the detailed canvassing of this sought after class of metastable alloys.

Recent work from our group has led us to an innovative synthetic approach in which we circumvent many of these issues. The method was first exemplified through the development of $AlPSi_3$ alloys integrated directly on Si(001) substrates (3). Decisive in the realization of this proof-of-concept material was the use of $P(SiH_3)_3$ as a growth precursor. The tailored growth chemistries employed interactions of Al atoms generated by an effusion cell and the $P(SiH_3)_3$ molecule, producing "$Al:P(SiH_3)_3$" intermediates. These intermediates subsequently eliminate terminal H as H_2, yielding a tetrahedron with fixed $AlPSi_3$ stoichiometry. These tetrahedral units interlink effectively fixing the local bonding environment through the incorporation of isolated Al-P dimers within the Si matrix. This synthetic strategy has all but eliminated the aforementioned hindrances.

This strategy has been fruitful and enabled conduct studies of the structure-property relationships in great detail. *En route* to understanding the optoelectronic properties of this new class of materials, we found that minor variations in the growth conditions can alter the nanoscale configurations of Al-P units within the alloy while maintaining the bulk $AlPSi_3$ stoichiometry (4). However this initial growth strategy had acute disadvantages that included fluctuations endemic to the Al beam during growth, and the limited production capacity afforded by the effusion source that severely limits the scalability of these materials into industrial processes. This prompted the pursuit of a an alternative Al source, for which $Al(BH_4)_3$ was chosen, to allay the problems of the effusion cell method. $Al(BH_4)_3$ is a carbon-free molecular source of Al that is chemically compatible with $P(SiH_3)_3$, and possesses a high enough vapor pressure at ambient temperatures (~150 Torr) to make it a viable precursor for CVD applications (5).

The decomposition of $Al(BH_4)_3$ proceeds through the elimination of diborane (B_2H_6) which is unreactive with $P(SiH_3)_3$ at typical reaction conditions. We anticipate the reaction between $Al(BH_4)_3$ and $P(SiH_3)_3$ will culminate in the formation of the desired "$H_3Al-P(SiH_3)_3$" intermediate, analogous to that formed in the pioneering work using the atomic Al effusion source. These building blocks eliminate terminal H atoms through the evolution of H_2 at the growth front generating the desired $AlPSi_3$ tetrahedral core. Other $M(TH_3)_3$ (M = P, As) (T = Si, Ge) molecular building blocks are combined with this novel Al source to demonstrate the robustness of this approach to other systems in the $(III-V)_{1-y}(IV_2)_y$ arena. The results presented in the subsequent text present an overview of recent progress which has been achieved in the systematic exploration of these materials and their recent extension to alloy systems fully lattice matched to Ge.

Materials Growth and Characterization of $Al_{1-x}B_xPSi_3$

The similarity in lattice parameter of $AlPSi_3$ (5.440 Å) and Si (5.431 Å) allowed for the pseudomorphic growth of thick layers, up to 900 nm, directly on Si platforms(3). However it may be possible to achieve full lattice matching through the incorporation of small amounts B by substitution for Al, mitigating any formation of strain related defects. Lattice matching is realized when the boron content x = 0.035, and though we don't expect B_2H_6 to function as an effective B source, the high concentration at the growth front may lead to ancillary incorporation. Additionally, modification of the lattice parameter is not anticipated to be the only boon with the addition of B. The incorporation of BP pairs is foreseen to increase the band gap of $AlPSi_3$ above the calculated 1.3 eV value (6), indicating that $Al_{1-x}B_xPSi_3$ materials may be ideally suited for photovoltaics where the ideal gap is 1.7 eV for a Si-based dual junction design (7).

<u>Growth of $Al_{1-x}B_xPSi_3/Si(001)$</u>

Silicon wafers were cleaved to 1.0 x 1.5 cm^2 rectangles and sonicated in 5% HF in methanol and rinsed in methanol before being dried under a stream of ultra-high purity N$_2$. The substrates were then loaded into the growth chamber and heated to 650° C to remove volatile impurities. Subsequently the substrates were subject to a flashing procedure in which they were heated to 1025° C for 10 seconds, ten times. To complete the surface preparation, a buffer layer of Si was deposited using Si$_4$H$_{10}$ diluted to 10% by volume in H$_2$ at T = 540° C and a pressure of 1x10^{-5} Torr. This produced 40-350 nm thick layers at a growth rate of 2 nm/min. Promptly after the deposition of the buffer layer, epilayers of the Al$_{1-x}$B$_x$PSi$_3$ alloy were grown between 535-545° C at 8.0-10x10^{-6} Torr using mixtures of Al(BH$_4$)$_3$ and P(SiH$_3$)$_3$ with a 1:2 molar ratio. Typical growth times were 15-60 minutes producing films ranging 70-570 nm at growth rates of 3-11 nm/min depending on temperature and pressure.

<u>Composition and Structure Determination</u>

Rutherford backscattering spectrometry (RBS) was used to determine the bulk compositions using an incident energy of 2.0 MeV in the Cornell geometry. The spectra show distinct signals from Al, P, and Si components in the films. The spectra were fitted using the RUMP package and indicate the resultant films have a nominal stoichiometry of AlPSi$_3$. Ion-channeling experiments suggest that the films are single crystal and are epitaxially aligned to the underlying Si substrates. Additionally RBS was collected at boron resonance energy, 3.9 MeV in an attempt to quantify boron uptake during the growth, however there is no detectable signal. This shows that the B content is below the detection limit of 5% for the technique.

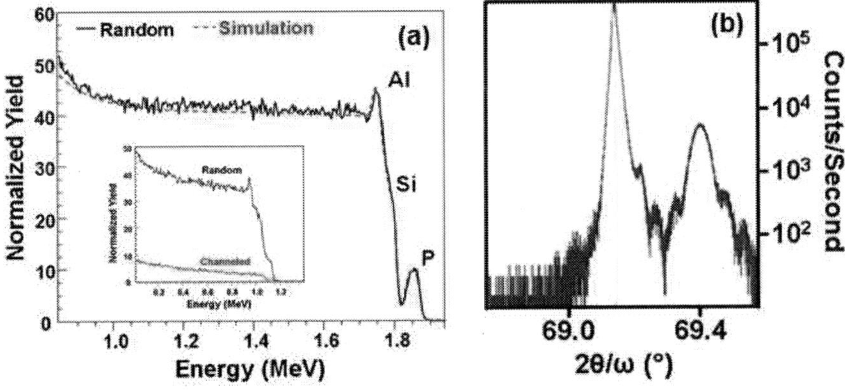

Figure 1. (a) Random 3.0 MeV RBS (solid) showing distinct signals for Al, P, and Si. Simulated spectrum (dashed) from RUMP yields a 180 nm thick film with composition AlPSi$_3$. The inset compares the random 2.0 MeV RBS to the (001) aligned spectrum, the high degree of channeling indicates epitaxial registry and the single crystal nature of the epilayer. (b) HR-XRD plot near the Si(004) reflection (main peak) showing an alloy peak at higher angular position with corresponding thickness fringes.

High-resolution X-Ray diffraction (HR-XRD) was used to determine lattice spacing and strain profiles of the alloys. HR-XRD measurements in the vicinity of the Si(004) reflection provided evidence of B incorporation. $2\theta/\omega$ plots contain a peak attributed to the alloy (004) reflection at a higher angular position relative to Si(004) indicating a smaller d-spacing due to appreciable B substitution in the parent $AlPSi_3$ lattice. Off-axis (224) HR-XRD plots were also collected to simultaneously determine $c_{\parallel}$ and $c_{\perp}$. The alloy and substrate peaks are aligned along the pseduomorphic line indicating $c_{\parallel}$ is identical to that of the Si substrate and the film is tetragonally distorted with $c_{\parallel} > c_{\perp}$.

<u>Microstructure and Analytical Electron Microscopy</u>

Cross-sectional scanning transmission electron microscopy (STEM) images affirm monocrystalline layers with uniform contrast. A JEOL ARM200F analytical microscope operated at 200 kV in STEM mode was used for imaging. The high-angle annular dark-field (HAADF) and the complimentary bright-field (BF) images show a typical sample in which the film is defect-free within the field of view and there are no evident signs of phase segregation.

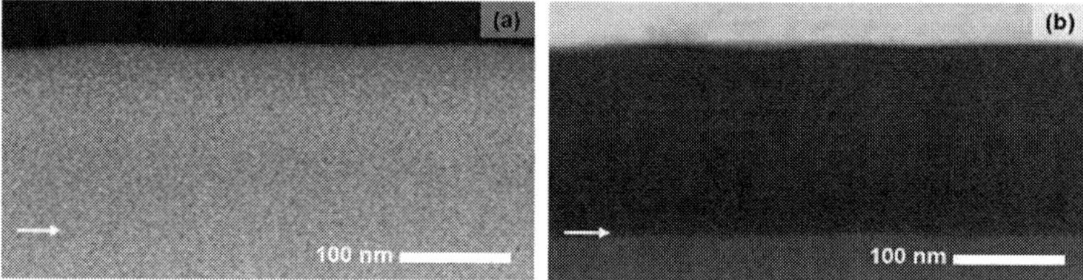

Figure 2. (a) Low magnification STEM HAADF image and (b) STEM bright field image showing the entire layer. The films have low defect density as there are no defects within the field of view, and they possess uniform contrast.

Bonding arrangements were examined using probe-corrected STEM-EELS performed on a NION UltraSTEM 100 operated at 100 kV. The structural and chemical analyses conducted in [110] projection indicate that the film is a single-phase alloy with homogeneous elemental distribution. Figure 3 shows atomic-resolution imaging and element-selective mapping. The maps were fitted using the Al K-, P K- and Si K-edges and all show distinct dimer-like features consistent with a diamond lattice. Additionally these maps indicate the Al-P pairs are randomly distributed in this projection, consistent with what is expected for a single-phase alloy with a disordered structure.

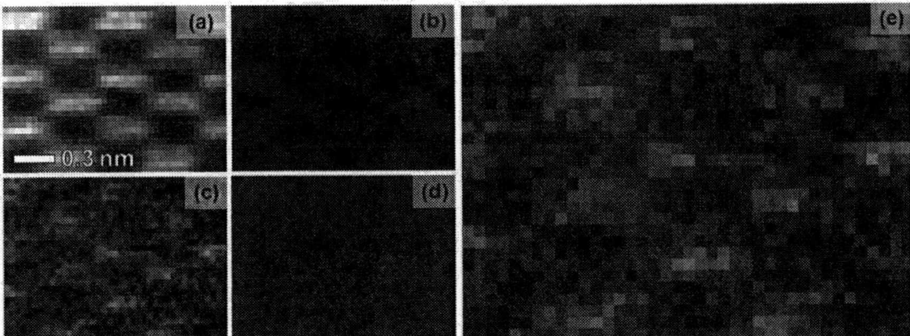

Figure 3. MLS fitted STEM-EELS map of $Al_{1-x}B_xPSi_3$ (a) HAADF image; (b) Si K-map; (c) Al K-map; (d) P K-map; (e) Enlarged color overlay of the Al, P, and Si maps.

Materials Growth and Characterization of $Al_{1-x}B_xAsT_3$ (T = Si, Ge)

The approach illustrated above grants a route to attain an analogous As containing alloy in which we consider reactions of $Al(BH_4)_3$ with $As(SiH_3)_3$ and $As(GeH_3)_3$. Experimentally we observe films with bulk compositions of $AlAsSi_3$ and $AlAsGe_3$ containing B at dopant concentrations. The significantly diminished incorporation of B into the $Al_{1-x}B_xAsSi_3$ alloys is attributed to the larger bond-strain associated with the formation of a B-As moiety compared to B-P in the previously described alloy system. The trend of exploring $(III-V)_{1-y}(IV_2)_y$ materials which are comprised of end members with similar lattice parameters is extended by the tetrahedral building block approach to reactions involving $As(GeH_3)_3$. The use of the $As(GeH_3)_3$ precursor has been implemented in conjunction with both atomic Al beams and the gaseous $Al(BH_4)_3$ source. The lattice parameter of AlAs is 5.660 Å and Ge is 5.658 Å, an alloy of these two end members in any concentration will be nearly lattice matched to a Ge substrate allowing for the facile integration on Ge platforms.

Growth and Characterization of $Al_{1-x}B_xAsSi_3$/Si(001)

The depositions of $Al(BH_4)_3$ and $As(SiH_3)_3$ were conducted at T = 580° C and 1.0×10^{-5} Torr directly on Si(001) substrates. The substrates were prepared using the same method described previously. The compositions were determined using RBS and were found to be $AlAsSi_3$ with thicknesses ranging 70-220 nm. The ion-channeling RBS was indicative of epitaxially aligned monocrystalline films on which the Al-As-Si components occupy a common lattice. HR-XRD of these films provided evidence that these films are compressively strained due the large lattice mismatch with Si. Off-axis HR-XRD combined with elasticity theory allowed us to determine a relaxed lattice parameter a_0 = 5.5217 Å, nearly the expected Vegard average and consistent with $AlAsSi_3$ films from prior work (8). Cross-sectional TEM micrographs corroborate the single-phase nature of this alloy system, but showed high defectivities in the form of dislocations and stacking faults distributed across the entire film. SIMS analysis was conducted to determine the level of B incorporation into the lattice, it was found to be 2-3×10^{20} cm^{-3}. *En masse* the these results demonstrate how apt $Al(BH_4)_3$ is as an effective molecular source of Al atoms, compatible with CVD processing.

Growth of $(AlAs)_{1-y}(Ge_2)_y$/Ge/Si(001)

The $(AlAs)_{1-y}(Ge_2)_y$ films were grown on Ge buffered Si(001) substrates by gas source molecular beam epitaxy (GS-MBE). The Ge buffered substrates were prepared using previously reported protocols in a gas source molecular epitaxy reactor using Ge_4H_{10} as a source material (9). The extremely high quality films are then used as templates. The as-grown wafers are cleaved into 1.0 x 1.5 cm^2 pieces which are compatible with the sample stage. The substrates are prepared via sonication in 5% HF/H_20 and methanol for 5 minutes each then dried under a steam of N_2. The substrates are then inserted into the chamber 0 a load-lock and then heated at 600° C to remove any residual volatile impurities. After the thermal cleaning procedure the substrate temperature was set to 525° C as measured by a single-color pyrometer. The depositions were initiated by first introducing Al atoms generated by the effusion cell to the chamber, then the gaseous $As(GeH_3)_3$ molecule was administered through a high-precision needle valve with a nozzle to produce a 1:1 molar ratio with the Al atoms. The system is allowed to equilibrate for a short period of time before the sample stage is moved into the growth position and the pressure was kept between $8.0\text{-}10x10^{-6}$ Torr. All of the depositions were conducted for 30 minutes, producing films ranging 350-565 nm at 12-19 nm/min dependent on growth pressure.

Growth of $(Al_{1-x}B_x)_{1-y}(Ge_2)_y$/Ge/Si(001)

Ge-on-Si substrates were made epi-ready using identical procedures to those for the antecedent section. The substrates were set to T = 500° C using a pyrometer. Mixtures of the gaseous precursors were combined in a 1 L glass bulb. The bulb was charged with a 1:2 molar ratio of $Al(BH_4)_3$:$As(GeH_3)_3$, and total pressures less than 1 Torr. The flux of precursor was introduced into the chamber at $1.0x10^{-5}$ Torr producing films up to 150 nm thick with growth times of up to 30 minutes and growth rates of ~5 nm/min.

Characterization of $(AlAs)_{1-y}(Ge_2)_y$ and $(Al_{1-x}B_xAs)_{1-y}(Ge_2)_y$ Films

For both of the alloy systems initial determination of bulk composition was done using RBS at 3.0 MeV to get the necessary mass resolution, typical spectra can be seen in Figure 4. In the case of the Al-As-Ge alloy system the Ge content never reached the ideal stoichiometric 1:3 As:Ge ratio, this is likely due to the unimolecular decomposition of the $As(GeH_3)_3$ precursor at the growth conditions. However in the case of the samples grown using $Al(BH_4)_3$ as the Al source, we found that the 1:3 ratio was preserved. In addition to compositional information the 2.0 MeV ion-channeling RBS data demonstrate that the films are epitaxially aligned and monocrystalline.

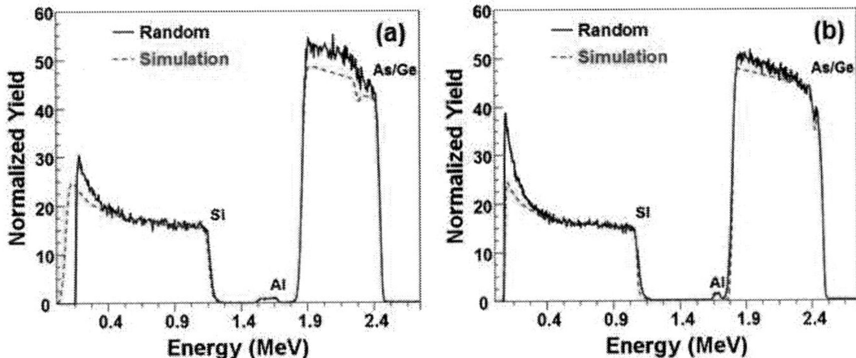

Figure 4. Representative 3.0 MeV RBS spectra for films grown on Ge/Si(001) using different Al sources. (a) Random RBS spectrum (solid) showing a unified edge for Ge and As, since their atomic numbers only differ by one, and a distinct signal for Al at lower energies. The RUMP model (dashed) corresponds to an alloy composition of AlAsGe$_7$ that is 350 nm thick grown using an Al effusion cell. (b) Random RBS (solid) for a typical sample grown using the Al(BH$_4$)$_3$ source. The RUMP model (dashed) corresponds to a 145 nm thick film with AlAsGe$_3$ composition.

As with the Al$_{1-x}$B$_x$PSi$_3$ samples the level of boron incorporation is below the limit of detection for RBS, SIMS analysis was conducted to determine the B concentration in the films grown using Al(BH$_4$)$_3$. The concentration of B was found to be 1.2-1.4x10^{20} cm^{-3} and all 4 elements are uniformly distributed through the layer.

Two representative samples, one grown using each approach was chosen and studied using a suite of electron microscopy techniques. Initial surveys of the microstructure were done using a JEOL ARM200F microscope operated in STEM mode at 200 kV. In this study we will refer to the samples as M, grown using the Al effusion cell, and C, grown using Al(BH$_4$)$_3$, the relevant parameters for these selected samples can be seen in Table I.

TABLE I. Growth data for samples M and C.

Sample	Compositions	Temperature	Pressure	Thickness
M	AlAsGe$_7$	527° C	1.0x10^{-5} Torr	350 nm
C	AlAsGe$_3$	501° C	1.0x10^{-5} Torr	145 nm

Low magnification HAADF imaging of both samples show modulation of the contrast. The variations are attributed to differences in the composition since the contrast is proportional to the atomic number of the constituent elements. Figure 5 shows a comparison of the two samples in low magnification, note the relative sizes of the domains are significantly different. In the case of sample M the variations appear to be striations propagating along the growth direction, but for sample C the modulations appear to be columnar in nature. The regions that are brighter have a higher average Z than the darker areas indicating they are Al-poor. To determine the origins of these contrast fluctuations and better understand the how each Al source effects the growth and subsequently the properties of the material, high-resolution STEM and small-probe analysis were conducted.

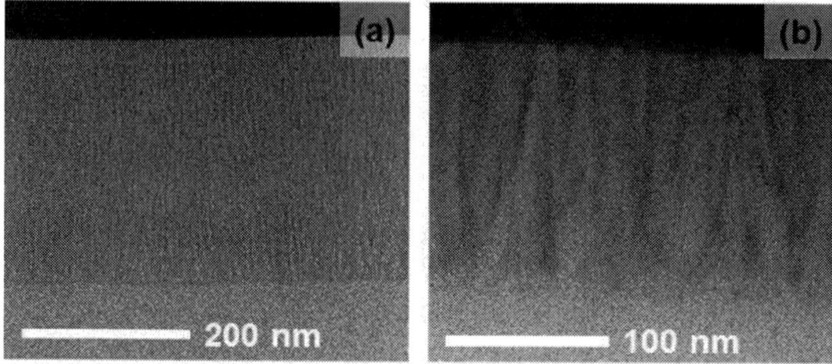

Figure 5. Low magnification HAADF images of samples M and C. (a) Cross sectional HAADF image of sample M grown using the solid Al source showing contrast modulation. (b) HAADF image of sample C grown using Al(BH$_4$)$_3$ showing a more pronounced modulation that hints there might compositional inhomogeneities.

A tableau of images can be seen in Figure 6 showing element-selective EELS mapping across several of these contrast variations seen in HAADF, for sample M, and the close correspondence with the compositional inhomogeneity shown in the color overlay.

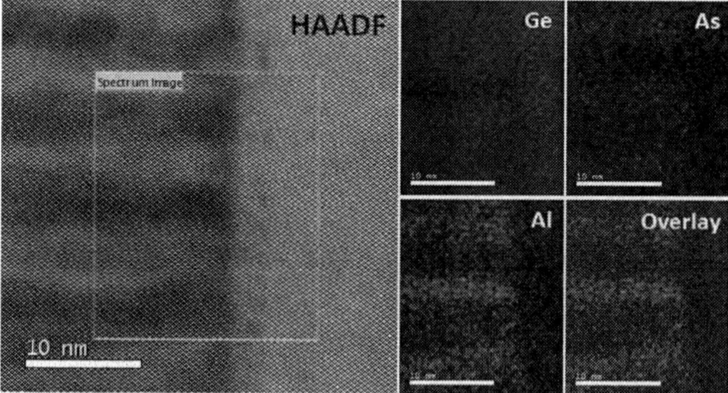

Figure 6. Medium magnification STEM-EELS analysis of sample M near the alloy-germanium interface; the left HAADF image shows the region which was mapped using EELS, the 2D EELS was collected within the box. MLS fits were performed on the EELS spectra to generate the Ge L-map (blue), As L-map (red) and Al K-map (green). The color overlay shows exemplifies the correspondence between the contrast modulations of the HAADF image with the observed compositional variations.

The EELS analysis was done using L-edges of Ge (1217 eV) and As (1323 eV) along with the K-edge of Al (1560 eV). The close proximity and delayed nature of these edges makes exact quantification difficult, inherently these analysis should be taken as very roughly quantitative. To corroborate the results obtained from the high-loss EELS mapping EDS mapping was concurrently done and yields similar qualitative results. The analysis that was conducted on sample M all indicate that there are compositional variations across the sample, but in every region all three elements are present. In contrast to the finding from sample M the EDS maps for sample C, shown in Figure 7 clearly indicate this is not the case for sample C.

Sample C was subjected to the same STEM-EELS and concurrent STEM-EDS using a small high current electron probe, below in Figure 7 we show a set of EDS maps that span several of the features seen in the low magnification images above. This sample which was grown using Al(BH$_4$)$_3$ as an Al source we can clearly see regions in which the Ge signal is absent. The regions in which this occurs are on the order of 5 nm, indicating that we have nanoscale phase segregation. High-resolution HAADF images were collected near the boundary of these chemically different regions, the results of which are complimentary to the analytical findings.

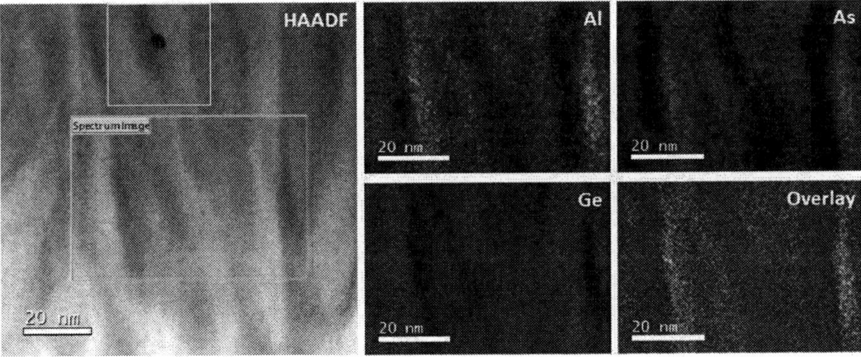

Figure 7. HAADF image (left) corresponding to the series of EDS elemental maps for Al (green), As (red), and Ge (blue). Note that the distributions of AlAs and Ge are complimentary, and the areas of dark contrast in HAADF correspond to regions which are significantly depleted of Ge.

The high-resolution HAADF images collected using the ARM200F allow us to clearly resolve the different atoms in the so-called dumbbells of a diamond like material in the [110] projection. The resolution afforded by probe-correction allowed us to see the asymmetric dumbbells present in the darker regions of the HAADF images. Figure 8(a) is a HAADF image clearly showing two different phases within the same film. The darker region in the middle was found to be Ge-poor through EELS and EDS analysis, this region is also distinctly lacking intensity on one half of the dumbbell, for a diamond-like material the dumbbells should be of equal intensity.

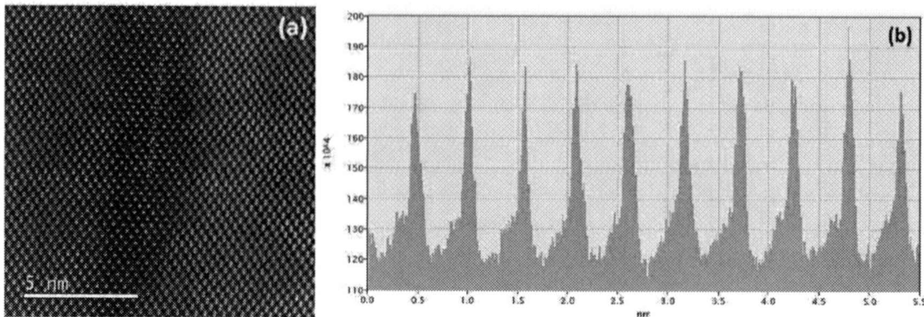

Figure 8. (a) High Resolution HAADF image showing the two distinct phases with nanoscale dimensions. The region in the center corresponds to one of the darker regions seen in low- and medium-magnification HAADF images (lower Ge content). A line profile 10 pixels wide was used to integrate the intensities, this plot is shown in (b). We can clearly see the dumbbells are complete but the intensity is diminished indicating the average Z within that column is significantly smaller than that of the neighboring column.

Conclusion

We have shown that $Al(BH_4)_3$ serves as an effective delivery source for Al in the growth of hybrid $(III-V)_{1-y}(IV_2)_y$ alloys, and is compatible with CVD processing protocols. This carbon-free source of Al atoms has been used to produce $Al_{1-x}B_xPSi_3$ and $Al_{1-x}B_xAsSi_3$ semiconductors with diamond-like cubic structures. In the cases described above, the reactions between the compound and $M(SiH_3)_3$ (M = P, As) produce single-phase layers composed of interlinked III-V-IV$_3$ tetrahedral units. As we speculated the incorporation of B was minimal due to the low reactivity of the B_2H_6 moiety at these growth conditions, but the collateral incorporation is expected to increase the gap of these materials as desired for tandem Si-based photovoltaic designs.

The elemental distribution was elucidated for two alloy systems, $(AlAs)_{1-y}(Ge_2)_y$ and $(Al_{1-x}B_xAs)_{1-y}(Ge_2)_y$ using aberration-corrected annular-dark-field imaging and concurrent chemical analysis using EDS and EELS. In both cases we see the formation of an epitaxially aligned film, but with compositional variations. The sample grown using an atomic beam of Al the inhomogeneities are likely due to fluctuations in the growth conditions such as temperature, pressure, and even the flux of precursor which may have an effect on instantaneous deposition profile. This finding is not unforeseen as the mechanism for which the assembly occurs is non-trivial. In the case of the sample grown using $Al(BH_4)_3$ we see a much more distinct variation of composition, to the point of phase segregation that was corroborated with EDS, EELS and high-resolution HAADF imaging. This result is not unsurprising as the growth was done using conditions more similar to CVD than MBE, at the temperatures required for the activation of these compounds the likelihood of deviation towards a more thermodynamically stable phase is greater. Through the judicious exploration of parameter space it might be possible to control the microstructure of these materials to produce a single-phase monocrystalline alloy with the desired composition.

Acknowledgments

We acknowledge the use of facilities with the LeRoy Eyring Center for Solid State Science at Arizona State University and John M. Cowley High Resolution Electron Microscopy at Arizona State University.

References

1. S. Chen, X. G. Gong, and S-H. Wei, *Phys. Rev. B.*, **77**, 014113 (2008).
2. A. G. Norman, J. M. Olson, J. F. Geisz, H. R., Moutinho, A. Mason, M. M. Al-Jassim, and S. M. Vernon, *Appl. Phys. Lett.*, **74**(10), 1382 (1999).
3. T. Watkins, A. V. G. Chizmeshya, L. Jiang, D. J. Smith, R. T. Beeler, G. Grzybowski, C. D. Poweleit, J. Menéndez, and J. Kouvetakis, *J. Am. Chem. Soc.*, **133**, 16212 (2011).
4. L. Jiang, T. Aoki, D. J. Smith, A. V. G. Chizmeshya, J. Menéndez, and J. Kouvetakis, *Chem. Mater.,* **26**, 4092 (2014).
5. H. I. Schlesinger, H. C. Brown, and E. K. Hyde, *J. Am. Chem. Soc.*, **75**, 209 (1953).
6. J.-H. Yang, Y. Zhai, H. Liu, H. Xiang, X. Gong, and S.-H. Wei, *J. Am. Chem. Soc.*, **134**, 12653 (2012).
7. J. F. Geisz and D. J. Freedman, *Semicond. Sci. Technol.*, **17**, 769 (2002).
8. G. Grzybowski, T. Watkins, R. T. Beeler, L. Jiang, D. J. Smith, A. V. G. Chizmeshya, J. Kouvetakis, and J. Menéndez, *Chem. Mater.*, **24**, 2347 (2012).
9. G. Grzybowski, L. Jiang, R. T. Beeler, T. Watkins, A. V. G. Chizmeshya, C. Xu, J. Menéndez, and J. Kouvetakis, *Chem. Mater.*, **24**, 1619 (2012).

Chapter 4

Nitride Materials & Devices

ECS Transactions, 69 (14) 97-102 (2015)
10.1149/06914.0097ecst ©The Electrochemical Society

From MRTA to SMRTA: Improvements in Activating Implanted Dopants in GaN

J.D. Greenlee[a], B.N. Feigelson[b],
T.J. Anderson[b], J.K. Hite[b], K.D. Hobart[b], F.J. Kub[b]

[a] NRC Postdoc Residing at the Naval Research Laboratory 4555 Overlook Ave SW,
Washington, DC 20375, USA
[b] Naval Research Laboratory 4555 Overlook Ave SW, Washington, DC 20375, USA

The implantation and activation of dopants in GaN is a key enabling step for future devices requiring selective area doping. High temperatures (over 1300 °C) are required to activate p-type dopants in GaN. These high temperatures are above the decomposition temperature of GaN, which necessitates advanced annealing processes. One such process, the multicycle rapid thermal annealing (MRTA) process, has enabled the implantation and activation of p-type dopants in GaN. In this research, we demonstrate the symmetrical multicycle rapid thermal annealing (SMRTA) annealing process. The SMRTA annealing process includes an extra conventional annealing step after the rapid heating and cooling cycles and is shown to improve the crystalline structure of annealed GaN compared to the MRTA process.

Introduction

GaN and related alloys have received a great deal of attention from the research community due to their tunable direct bandgap, radiation hardness, and a favorable Baliga figure of merit compared to SiC and Si (1, 2). Many innovative devices have been enabled by GaN and its related alloys, including blue light-emitting diodes and microwave power amplifiers (3). However, processing challenges for GaN still remain. One of these challenges is the ability to selectively implant and activate p-type dopants. Devices that will be enabled by the ability to selectively dope GaN include the current aperture vertical electron transistor (CAVET), the vertical GaN-based trench gate metal oxide semiconductor FET, and the junction gate field-effect transistor, (4-7). Unfortunately, the temperatures required to activate implanted p-type dopants, such as Mg, are higher than that of the decomposition temperature of GaN. Activation of implanted Mg requires temperatures above 1300 °C while GaN decomposes at around 840 °C at atmospheric pressures of N_2 (8).

Several methods have been employed to anneal GaN without decomposition. One method is to use a significant N_2 overpressure. To achieve temperatures up to 1500 °C without decomposition requires a 1.5 GPa N_2 overpressure (9). However, this large overpressure is not easily scalable and requires a complicated experimental setup.

The MRTA Technique

A more scalable method of annealing GaN while avoiding decomposition includes non-equilibrium annealing. A non-equilibrium annealing process has been previously implemented with the multicycle rapid thermal annealing (MRTA) process, which has been applied to SiC and GaN (10, 11). The MRTA process temperature treatment consists of two subsequent steps, a conventional anneal followed by rapid heating/cooling cycles (10). The conventional anneal is performed at temperatures where GaN is stable and is utilized to repair some of the damage induced during implantation. This conventional annealing step has been optimized based on crystal quality and surface morphology considerations (12). It was determined that higher conventional annealing temperatures result in better crystalline quality, but temperatures above 1150 °C resulted in surface decomposition and a roughened surface morphology (12). After the conventional annealing step, rapid heating and cooling pulses are applied to the GaN to activate the implanted dopants. Annealing Mg-implanted GaN with this MRTA process has resulted in p-type activation with efficiencies up to 8% (13).

Experimental Design – The SMRTA Technique

Although the MRTA process has demonstrated p-type activation in GaN, a modified annealing process is required to improve the reliability and consistency of the annealed samples. It has been previously determined that conventional annealing can remove much of the implantation induced damage in GaN (12). Thus, it was hypothesized that by adding an additional conventional annealing step at the end of the process, any detrimental structural changes caused by the rapid heating and cooling cycles can be reversed and an improved annealing process will result. The improved annealing process, termed symmetrical multicycle rapid thermal annealing (SMRTA), includes an additional annealing step at the end of the MRTA process as shown in Figure 1.

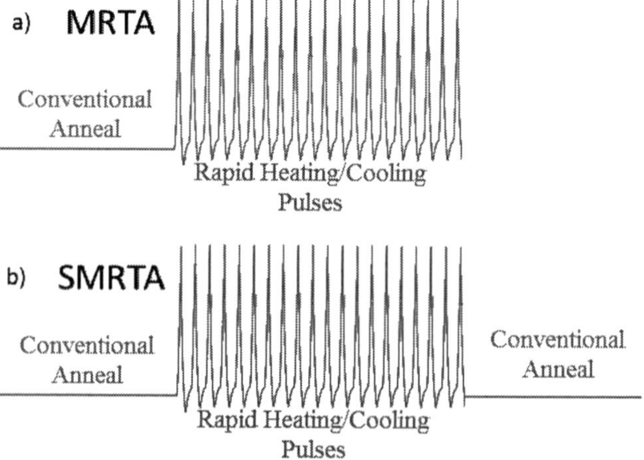

Figure 1. MRTA (a) and SMRTA (b) annealing temperature profile schematics. The MRTA process includes a preliminary conventional anneal followed by rapid heating/cooling pulses. The SMRTA process follows the same annealing procedure as the MRTA process, but includes an additional conventional anneal after the rapid heating/cooling pulses.

In this research, the crystal quality and surface morphology of GaN thin films annealed with the MRTA and SMRTA processes are investigated and compared. The two samples used in this experiment were grown on a sapphire substrate by metal organic chemical vapor deposition (MOCVD) and consisted of 2 μm of GaN with a three-layer AlN capping structure. The first thin AlN layer (4nm) was grown at high temperature (1100 °C) for an improved interface without cracking (10). The second AlN layer (25nm) was grown at 660 °C, and was included for mechanical support. The third layer is a 250 nm thick AlN layer, which was sputtered on the surface of the MOCVD grown AlN stack to provide additional mechanical support. This additional sputtered cap provided further mechanical support and has been previously shown to improve the resilience of the MOCVD AlN cap (14). The samples were masked using photoresist such that half of each sample remained unimplanted. The implantation was conducted at room temperature with the doses and energies shown in Table 1. These energies and doses were chosen for a constant Mg concentration of approximately 1×10^{19} cm^{-3}.

TABLE I. Implantation energies and doses used in this research. The energies and doses were chosen for a constant Mg concentration of approximately 1×10^{19} cm^{-3} up to a depth of 500 nm.

Energy (keV)	Dose (atoms/cm^2)
50	5.0×10^{13}
140	1.1×10^{14}
300	3.0×10^{14}

One of the samples studied was annealed with the MRTA process and the other was annealed with the SMRTA process. The MRTA process included a conventional anneal at 1000 °C for 30 minutes followed by 40 rapid heating and cooling cycles with a peak temperature of 1350 °C. The SMRTA process had the same conventional anneal and pulse conditions, but the process was followed by an additional conventional anneal at 1000 °C for 30 minutes. The samples were characterized using Raman spectroscopy, photoluminescence spectroscopy (PL), and Nomarski microscopy.

Results and Discussion

Raman Spectroscopy has been used previously to characterize implanted GaN (15). The full width at half-maximum (FWHM) of the GaN E_2 mode is commonly used to provide crystal quality information, where a lower FWHM corresponds with an improvement in crystalline quality. Before and after the MRTA and SMRTA annealing processes detailed above, the E_2 mode was monitored to quantify changes in the crystalline quality. As implanted, the E_2 mode is severely broadened compared to the as-grown portion of the sample due to implantation damage. After the MRTA annealing process, the E_2 FWHM dramatically decreases due to the removal of some of the implantation damage. However, the as-grown FWHM increases, indicating that the MRTA process degraded the GaN crystal quality. The SMRTA process results in a decrease in the FWHM compared to the MRTA process for both the unimplanted and implanted GaN, indicating that the conventional anneal at the end of the SMRTA process is important for removing the crystalline degradation caused by the rapid heating and cooling cycles in the MRTA process.

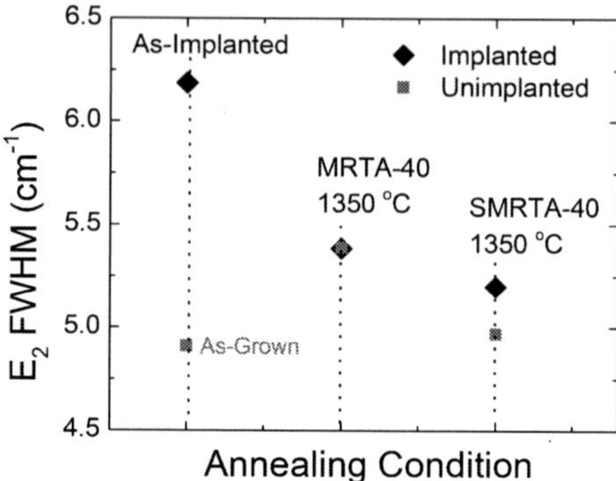

Figure 2. Raman E_2 FWHM with different annealing conditions. After the initial implantation, the FWHM increases dramatically compared to the as-grown portion of the sample. The sample annealed with the SMRTA process results in an improvement in both the implanted and unimplanted portions over the MRTA-annealed sample.

This improvement in crystalline quality was also confirmed with photoluminescence spectroscopy. In Mg-implanted GaN, the yellow PL emission has been used as an indication of recovery from the implantation damage (16). As shown in Figure 3, the yellow emission compared the band edge emission is lower after the SMRTA annealing process than the MRTA process, confirming the crystalline quality improvement indicated by the Raman data in Figure 2.

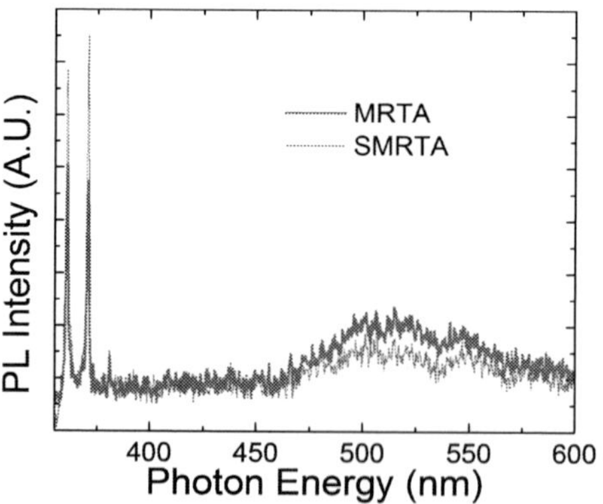

Figure 3. PL spectra of the implanted regions annealed with the SMRTA and MRTA processes. The ratio of the band edge luminescence compared to the yellow luminescence is higher for the SMRTA annealed sample, indicating an improvement in crystal quality.

The surface morphology of the SMRTA annealed sample was also investigated using Nomarski microscopy. Before annealing, the sample is smooth with no macroscopic defects. After annealing, the sample morphology does not change as shown in Figure 4. No cracking or macroscopic damage associated with GaN decomposition was observed. Thus, the capping structure is effective in protecting the GaN during the annealing process.

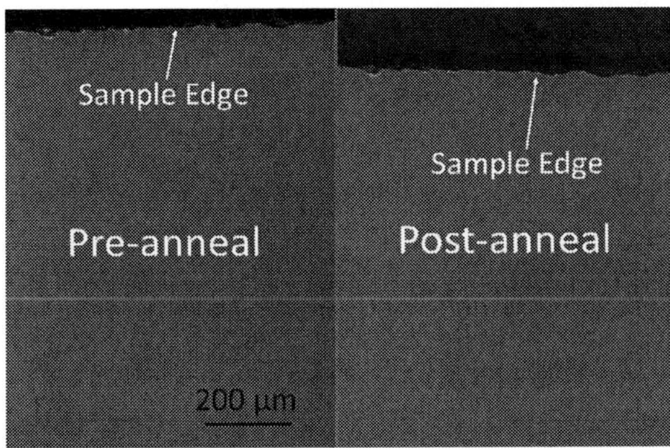

Figure 4. Nomarski microscopy image of the sample surface with the AlN cap still in place before (left) and after (right) annealing. After annealing, the sample surface remains smooth.

Conclusion

A novel annealing process, SMRTA, has been developed. The SMRTA process differs from the previous MRTA process in that it includes an additional conventional annealing step after the rapid heating and cooling cycles. It was shown with Raman spectroscopy and PL that the crystalline quality of SMRTA annealed implanted GaN is improved over that annealed with the MRTA process. The SMRTA process, demonstrated herein, will be a key enabling step for future transformative GaN-based devices requiring selective area doping.

Acknowledgments

This research was performed while J. D. Greenlee held a National Research Council Research Associateship Award at the Naval Research Laboratory. Research at NRL was supported by the Office of Naval Research.

References

1.   T. P. Chow and R. Tyagi, *IEEE Trans. Elec. Dev.*, **41**, 1481 (1994).
2.   T. J. Anderson, A. D. Koehler, J. D. Greenlee, B. D. Weaver, M. A. Mastro, J. K. Hite, J. C. R. Eddy, F. J. Kub and K. D. Hobart, *IEEE Electron Device Lett.*, **35**, 826 (2014).

3.   S. Nakamura, T. Mukai and M. Senoh, *Jpn. J. Appl. Phys.*, **30**, L1998 (1991).
4.   T. P. Chow, *Microelectron. Eng.*, **83**, 112 (2006).
5.   S. Chowdhury, B. L. Swenson and U. K. Mishra, *IEEE Electr. Device Lett.*, **29**, 543 (2008).
6.   O. Tohru, U. Yukihisa, I. Tsutomu and H. Kazuya, *Appl. Phys. Express*, **7**, 021002 (2014).
7.   J. C. Zolper, R. J. Shul, A. G. Baca, R. G. Wilson, S. J. Pearton and R. A. Stall, *Appl. Phys. Lett.*, **68**, 2273 (1996).
8.   J. Unland, B. Onderka, A. Davydov and R. Schmid-Fetzer, *J. Cryst. Growth*, **256**, 33 (2003).
9.   M. Kuball, J. M. Hayes, T. Suski, J. Jun, M. Leszczynski, J. Domagala, H. H. Tan, J. S. Williams and C. Jagadish, *J. Appl. Phys.*, **87**, 2736 (2000).
10.   B. N. Feigelson, T. J. Anderson, M. Abraham, J. A. Freitas, J. K. Hite, C. R. Eddy and F. J. Kub, *J. Cryst. Growth*, **350**, 21 (2012).
11.   M. J. Tadjer, N. A. Mahadik, B. N. Feigelson, R. E. Stahlbush, E. A. Imhoff, P. B. Klein, J. A. Freitas, J. D. Greenlee and F. J. Kub, in *Materials Science Forum*, p. 297 (2015).
12.   J. D. Greenlee, B. N. Feigelson, T. J. Anderson, M. J. Tadjer, J. K. Hite, M. A. Mastro, C. R. Eddy, K. D. Hobart and F. J. Kub, *J. Appl. Phys.*, **116** (2014).
13.   T. J. Anderson, B. N. Feigelson, F. J. Kub, M. J. Tadjer, K. D. Hobart, M. A. Mastro, J. K. Hite and C. R. E. Jr., *Electronics Lett.*, **50**, 197 (2014).
14.   J. D. Greenlee, T. J. Anderson, B. N. Feigelson, J. K. Hite, K. M. Bussmann, J. Charles R. Eddy, K. D. Hobart and F. J. Kub, *Appl. Phys. Express*, **7**, 121003 (2014).
15.   M. Katsikini, K. Papagelis, E. C. Paloura and S. Ves, *J. Appl. Phys.*, **94**, 4389 (2003).
16.   S. Hashimoto, T. Nakamura, Y. Honda and H. Amano, *Journal of Crystal Growth*, **388**, 112 (2014).

ECS Transactions, 69 (14) 103-108 (2015)
10.1149/06914.0103ecst ©The Electrochemical Society

Study of the effects of GaN buffer layer quality on the dc characteristics of AlGaN/GaN high electron mobility transistors

Shihyun Ahn[a], Weidi Zhu[a], Chen Dong[a], Ya-Hsi Hwang[a], Byung-Jae Kim[a], Fan Ren[a], Stephen J. Pearton[b], Aaron G. Lind[b], Kevin S. Jones[b] and Ivan I. Kravchenko[c]

[a] Department of Chemical Engineering, University of Florida, Gainesville, Florida 32611
[b] Department of Material Science and Engineering, University of Florida, Gainesville, Florida 32611
[c] Center for Nanophase Materials Sciences, Oak Ridge National Laboratory, Oak Ridge, Tennessee 37830

The influence of different buffer layer quality on AlGaN/GaN high electron mobility transistors (HEMTs) dc characteristics was investigated. Dc measurement by parameter analyzer and gate pulse measurement were performed to compare two different buffer layer quality samples. The same Al concentrations of AlGaN with 2µm and 5µm GaN buffer layers on sapphire substrates from two different vendors were used. The defect densities of 2 µm and 5 µm GaN buffer layer's HEMTs structures measured by transmission electron microscopy (TEM) were 7×10^9 cm^{-2} and 5×10^8 cm^{-2}, respectively.
There was small difference in drain saturation current and transfer characteristics in HMETs for these two types of buffer. Non-passivated HEMT with 5 µm GaN buffer layer showed no dispersion in gate-lag pulsed measurement at 100 kHz but HEMT with 2 µm GaN buffer layer showed 71% drain current reduction at 100 kHz gate-lag pulsed measurement.

Introduction

AlGaN/GaN high electron mobility transistors (HEMTs) are promising candidates for high power and high current applications in advanced radar systems, inverter units in hybrid electric vehicles, space and satellite communication networks with its large bandgap in the AlGaN/GaN heterostructure as well as the high electron mobilities and breakdown fields[1-7]. It is well known to have dislocation densities in the order of 10^8-10^{10} cm^{-2} for epitaxial films growth on sapphire or SiC substrates due to their thermal coefficient and lattice mismatches [8-10]. With its superior thermal conductivity, SiC is an excellent substrate candidate for power applications, however, the cost of SiC substrates is high. Therefore, sapphire is a promising candidate with its advantage in lower price and often superior surface quality despite its low thermal conductivity.

Optimizing GaN buffer layer growing conditions to reduce defect densities have been well studied[11-13]. Thinner GaN buffer layers on SiC have been reported to have higher threading dislocation density in the GaN with higher off-state breakdown voltage[12,13]. However, impact of thick GaN buffer layers on sapphire substrate on dc performance has not been widely studied. In this work, the effect of GaN buffer layer quality on AlGaN/GaN HEMTs' dc and gate-lag pulse performance was investigated. TEM was used to evaluate the dislocation density in the GaN buffer layer of the HEMT samples. Drain I-Vs, transfer characteristics and gate pulsed drain I-V characteristics of the HEMTs were used to find a correlation between the AlGaN/GaN HEMTs' dc and gate-pulsed performance and GaN buffer layers quality.

Experimental

Metal organic chemical vapor deposition (MOCVD) grown AlGaN/GaN HEMT structures on sapphire substrates were acquired from two different vendors. AlGaN barrier layer was consisted of 24% Al concentration. One wafer had a 5 μm GaN buffer and the other had a 2 μm GaN buffer layer. The mesa-etching was done by using a Unaxis Shuttle-lock Reactive Ion Etcher with Inductively Coupled Plasma Module (ICP) to attain device isolation with a Cl_2/Ar plasma followed by conventional photolithography. Electron-beam evaporator was used for Ohmic metallization of Ti/Al/Ni/Au (25nm/125nm/45nm/100nm) and subsequent rapid thermal annealing was done at 850°C in a flowing N_2 ambient for 45seconds. A Schottky gate was deposited by e-beam evaporated Ni/Au (20nm/80nm) with 100 μm width and 1 μm length. HP4156 parameter analyzer was used to obtain dc current-voltage (I-V) measurement and Agilent 8114A pulse generator was used to acquire gate-lag pulse measurement.

Result and Discussion

Figure 1 shows the cross-sectional transmission electron microscopy (XTEM) images of two different HEMT wafers. Figure 1(a) shows AlGaN/GaN HEMT's top surface structure with 2 μm GaN buffer layer, (b) shows the HEMT structure with 5 μm GaN buffer layer, (c) and (d) show the interface of 2 and 5 μm AlGaN/GaN HETMs' GaN buffer layers to sapphire substrates, respectively. Threading dislocations are shown to originate from the interface between GaN buffer layers and sapphire substrate to the HEMTs' surface for both wafers. It is noticeable that the higher defect densities are observed for the wafer with thinner GaN buffer layer. The measured defect densities, which are from bright-field cross-sectional TEM (BF-XTEM) taken along the length of the TEM lamella measured during lamella formation with a focused ion beam (FIB), were 5×10^8 cm^{-2} and 7×10^9 cm^{-2} for 5 μm and 2 μm GaN buffer layer substrate HEMTs respectively. One order of magnitude higher defect densities were present in the thinner GaN buffer layer HEMT wafer.

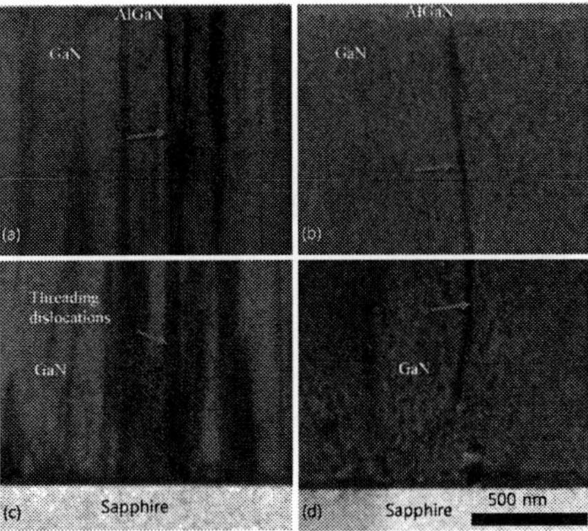

Figure 1. Bright Field cross-section TEM image of AlGaN/GaN interface for the (a) 2 μm GaN and (b) 5 μm GaN substrates and the GaN/sapphire interface for the (c) 2 μm GaN and (d) 5 μm GaN substrates

Despite its less defect density, 5 μm GaN buffer layer HEMT shows that it was under a high strain. Figure 2 shows the HEMT wafer with 5 μm GaN buffer layer was significantly bowed compare to the HEMT wafer with 2 μm GaN buffer layer. 5 meter radius of curvature was calculated for HEMT with 5 μm GaN buffer layer, which had a height difference of 70 μm across the 2" wafer. In the other hand, 45 meter radius of curvature was calculated for HEMT with 2 μm GaN buffer layer, which had a height difference of 21.25 μm across the 3" wafer. These show that the HEMT wafer with 5 μm buffer layer was highly strained.

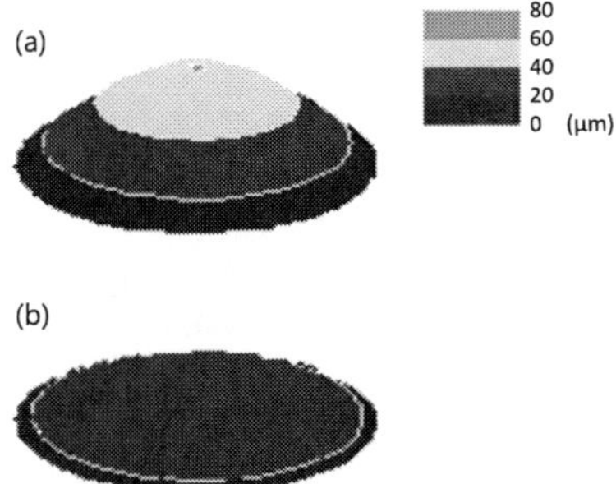

Figure 2. HEMT wafer's schematics of wafer bowing for with (a) 5 μm and (b) 2 μm GaN buffer layer

The drain current-voltage (I-V) characteristics of HEMTs fabricated on the different thickness of buffer layers were shown in figure 3(a). V_{DS} was swept from 0V to 5V while V_G started from +1V with -1V step. Little difference in the saturation current at V_{DS} = +5V. Figure 3(b) shows the transfer characteristics of the two different HEMTs. HEMT with thicker GaN buffer layer showed slightly higher saturation current at V_G = 0V with current density of 312 mA/mm whereas the HEMT with thinner GaN buffer layer was 300 mA/mm. The peak transconductance for

HEMT with 5 µm GaN buffer was slightly higher than HETM with 2 µm GaN buffer layer, which were 119 mS/mm and 112 mS/mm respectively. The mobility and carrier concentration were calculated using charge control model[14,15] on linear region of their drain I-V curve. The calculated mobility and carrier concentrations were 989 cm^2/V·S and 8.45 x10^{12} cm^{-2} for thicker GaN layer HEMT and 907 cm^2/V·S and 8.63 x 10^{12} cm^{-2} for thinner GaN layer HEMTs. The small differences in the dc measurements suggest that the defect density in HEMT's GaN buffer layer has small impact on their dc performance.

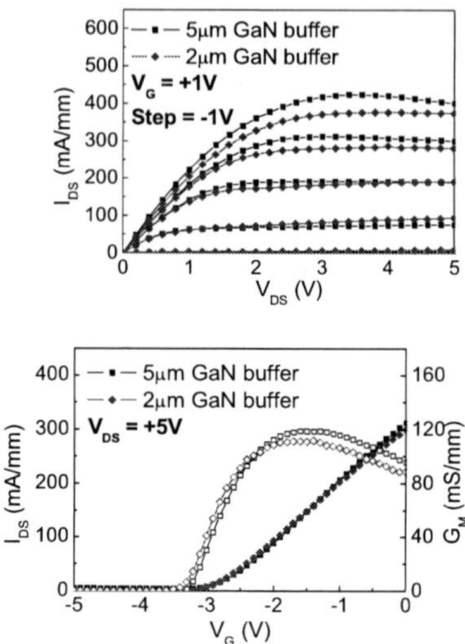

Figure 3. Different GaN buffer layer structure HEMTs' (a) drain I-Vs and (b) transfer characteristics

Material's electrical quality was evaluated using gate-lag pulse measurement. The drain current was measured while the gate voltage was pulsed. Figure 5 show the normalized drain current, pulsed-gate drain current versus gate voltage. Gate voltage was swept from -5V to 0V for normalized drain current and pulsed from -5V to the voltage indicated on the x-axis in figure 5 at 100 kHz. Figure 5(a) shows a no reduction of drain current at pulsed V_G = 0V. In the other hand, figure 5(b) shows dramatic reduction in the drain current at pulsed V_G = 0V. When the gate is pulsed at certain high frequency, the hot electrons are trapped in the surface traps and act as a virtual gate, reducing the drain current[15]. The 71% of drain current reduction in the gate pulsed measurement for the HEMT with 2 µm GaN buffer layer at V_G = 0V indicates that the excess number of defects in the thinner GaN buffer layer influenced the generation of hot electrons getting trapped in surface trap and caused drain current collapse at high frequency operation.

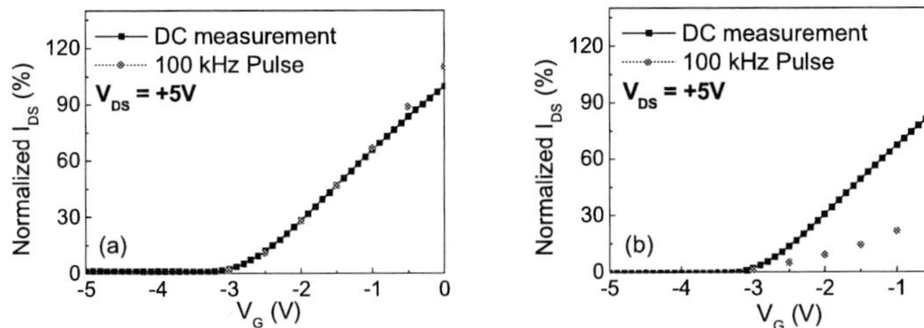

Figure 4. HEMT's gate-lag measurement for (a) 5 μm GaN buffer layer sample and (b) 2 μm GaN buffer layer

Conclusion

The impact of GaN buffer layer's defect density on HEMT dc and rf performance was studied. Thicker GaN buffer layer wafer showed one order higher defect density but had higher strain in the HEMT wafer. However, small difference was seen on their drain I-V characteristics with slightly higher in peak transconductance. In the other hand, significant reduction in gate-lag measurement was shown in the HEMT with thinner GaN buffer layer. The results suggest that the defects generated in GaN buffer layer have small influence in the device' dc performance but has significant impact on the high frequency performance.

Acknowledgments

The work performed at UF is supported by an U.S. DOD HDTRA Grant No. 1-11-1-0020 monitored by Dr. James Reed and a NSF Grant No. ECCS-14457 20 monitored by John Zavada. Authors would like to thank Ajit Paranjpe for materials support. A portion of this research was conducted at the Center for Nanophase Materials Sciences, which is a DOE Office of Science User Facility.

References

[1] P. Srivastava, J. Das, D. Visalli, M. V. Hove, P. E. Malinowski, D. Marcon, S. Lenci, K. Geens, K. Cheng, M. Leys, S. Decoutere, R. P. Mertens and G. Borghs, IEEE Electron Dev. Lett. 32, 30 (2011).
[2] H. Sun, A. Alt, H. Benedickter, and C. Bolognesi, IEEE Electron Dev. Lett. 45, 376 (2009).
[3] L. Bin and T. Palacios. IEEE Electron Dev. Lett. 31, 9 (2010).
[4] M. Ochiai, M. Akitai, Y. Ohno, S. Kishimoto, K. Maezawa, and T. Mizutani, Jpn. J. Appl. Phys., 42, 2278 (2003).
[5] T. Hashizume, S. Anantathanasarn, N. Negoro, E. Sano, H. Hasegawa, K. Kumakura, and T. Makimoto, Jpn. J. Appl. Phys., 43, L777 (2004).
[6] M. Kanamura, T. Ohki, T. Kikkawa, K. Imanishi, T. Imada, A. Yamada, and N. Hara, IEEE Electron Dev.. Lett. 31, 189 (2010).

[7]Y. Irokawa, Y. Nakano, M. Ishiko, T. Kachi, J. Kim, F. Ren, B.P. Gila, A. Onstine, C.R. Abernathy, S.J. Pearton, C.C. Pan, G.T. Chen, and J.I. Chyi, Appl. Phys. Lett. 84, 2919 (2004).

[8]M. A. Khan, X. Hu, A. Tarakji, G. Simin, J. Yang, R. Gaska, and M.S. Shur, Appl. Phys. Lett. 77, 1339 (2000).

[9]A. Hinoki, J. Kikawa, T. Yamada, T. Tsuchiya, S. Kamiya, M. Kurouchi, K. Kosaka, T. Araki, A. Suzuki, and Y. Nanishi, Appl. Phys. Express. 1, 011103 (2008).

[10]A.E. Wickenden, D.D. Koleske, R.L. Henry, M.E. Twigg, and M. Fatemi, Cryst. Growth 260, 54 (2003).

[11]Z. Bougrioua, i. Moerman, L. Nistor, B. Van Daele, E. Monroy, T. Palacios, F. Calle, and M. Leroux, Phys. Status Solidi (a) 195, 93 (2003).

[12]A. Hinoki, S. Kamiya, T. Tsuchiya, T. Yamada, J. Kikawa, T. Araki, A. Suzuki, and Y. Nanishi, Phys. Status Solidi (c) 4, 2728 (2007).

[13]L. Liu, C.F. Lo, Y. Xi, F. Ren, S. J. Pearton, O. Laboutin, Y. Cao, W. Johnson, I.I. Kravchenko, J. Vac. Sci. Technol. B 31, 011805 (2013).

[14]Y. Zhang, I.P. Smorchkova, C.R. Elass, S. Keller, J.P. Ibbertson, S. DenBaars, U.K. Mishra, and J. Singh, J. Appl, Phys. 87, 7981 (2000)

[15]R. Vetury, N. Q. Zhang, S. Keller, and U. K. Mishra, IEEE Trans. Electron Dev. 48, 560 (2001).

Chapter 5

Poster Session

ECS Transactions, 69 (14) 111-118 (2015)
10.1149/06914.0111ecst ©The Electrochemical Society

Effect of Buffer Oxide Etchant (BOE) on Ti/Al/Ni/Au Ohmic contacts for AlGaN/GaN based HEMT

Ya-Hsi Hwang[a], Shihyun Ahn[a], Chen Dong[a], Weidi Zhu[a], Byung-Jae Kim[a], Fan Ren[a], Aaron G. Lind[b], Kevin S. Jones[b], Stephen J. Pearton[b] and Ivan I. Kravchenko[c]

[a] Department of Chemical Engineering, University of Florida, Gainesville, Florida, 32611, USA
[b] Department of Materials Science and Engineering, University of Florida, Gainesville, Florida, 32611, USA
[c] Center for Nanophase Materials Sciences, Oak Ridge National Laboratory, Oak Ridge, Tennessee 37830

The effects of buffer oxide etchant on Ti/Al/Ni/Au Ohmic contacts for AlGaN/GaN high electron mobility transistor were studied. Scanning electron microscopy (SEM), energy dispersive spectrum (EDX), and Auger electron spectroscopy (AES) were performed to investigate the degradation after treatment. Contact resistance increase from 1×10^{-4} to 2.5×10^{-4} ohm/cm^2 after treatment. Ohmic metals were found out to be divided into three regions including islands, rings surround the islands, and the remaining field area. The islands were found to be shorter and the rings thinner after treatment. Besides, Ti, Al and Ni were found to be etched by BOE and caused Au to peel off from the Ohmic metals. The etching of Ti, Al and Ni was believed to be the reason of degradation.

Introduction

AlGaN/GaN high electron mobility transistors (HEMTs) have received increasing attention for high power and high frequency applications such as military radar and satellite-based communications systems due to their superior mobility (~1400 cm^2/V-s) and larger energy bandgap as compared to Si-based power transistors [1,2]. The qualities of Ohmic contacts are key characteristics to guarantee the high mobility. Among the Ohmic contacts, Ti/Al/Ni/Au is the most common metal stack for AlGaN/GaN HEMT. There are several studies about improving the Ohmic performances[3], changing the metal scheme to lower down the thermal budget [4], but few of them discuss the reliability of Ohmic contacts.

During fabrication, Ohmic metals would be exposed to chemicals such as Buffer Oxide Etchant (BOE), and development solution. BOE is a mixture of HF and NH$_4$F, which is known to be reactive with metals such as Ti, Al, Ni spontaneously [5,6]. As a result, it is necessary to understand the effect of BOE treatment on Ohmic metals. In this work, a systematic study of the effect of BOE to Ohmic contacts was performed. Transmission line method (TLM) was used first to characterize the performance of Ohmic contacts before and after BOE treatment. Optical microscopy and scanning electron microscopy (SEM) were performed to observe the surface of Ohmic metals before and after treatment.

Last, energy dispersive spectrum (EDX) and Auger electron microscopy (AES) was demonstrated to characterize the composition change before and after BOE treatment.

Experimental

The HEMT structures consisted of a 25 nm AlGaN barrier layer, a 0.8 μm C doped GaN buffer layer, a 1.4 μm AlGaN transition layer and a thin AlN nucleation layer on 300 μm Si substrate. Ti/Al/Ni/Au (45 nm/125 nm/100 nm/45 nm) was deposited by electron beam evaporation and standard lift-off process. The Ohmic metals were subsequently annealed in N_2 at 850°C for 1 min. Ohmic metals were exposed to BOE for 3 mins. The contact resistance was monitored every 15 sec by transmission line method (TLM). Scanning electron microscopy (SEM) was used to observe the surface morphology change after BOE exposure. Energy dispersive X-ray spectrum (EDX) and Auger electron analysis (AES) was used to characterize the elemental change after BOE exposure.

Results and Discussion

Figure 1 shows the effect of BOE exposure to contact resistance. As shown in Figure 1, the contact resistance increased from 1×10^{-4} Ω-cm^2 to 2.5×10^{-4} Ω-cm^2 after exposure to BOE for 2 minute. BOE solution is a solution which composed of NH_4F and HF. Since the reaction of Ti, Al, and Ni to HF is spontaneous, the increase of metal resistance is expected. The sheet resistance doesn't change much with BOE treatment time, indicating the GaN is more inert to BOE exposure. In other words, the increase of contact resistance is mainly due to the increase of metal resistance from exposure to BOE.

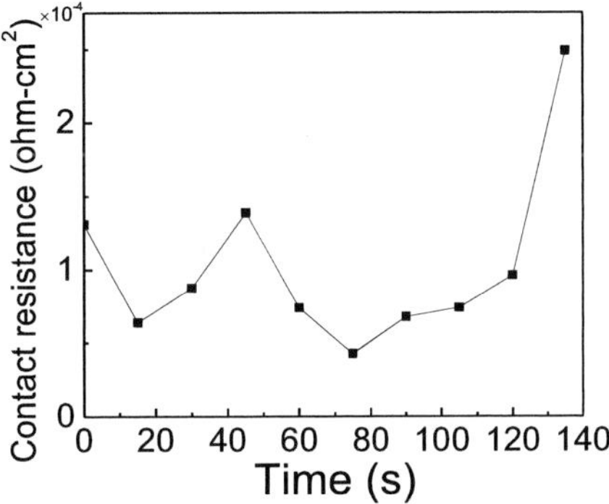

Figure 1. Effect of BOE treatment time to contact resistance

To further observe the surface morphology of Ohmic contacts before and after BOE treatment, scanning electron microscope (SEM) was utilized. As shown in Figure 2, Ohmic contacts can be furthered into three regions, i.e. islands, rings, and field areas. Islands are formed during annealing to decrease the interfacial energies based on the previous report.[6] The size of the ring depends on the annealing temperature, which is usually 5μm at 850°C. Rings, which surround the island, are usually 1 μm wide and are composed of Au. After BOE treatment, the size and the density of the islands doesn't change much. However, the island becomes thinner and there is a trench formed outside the ring. To understand the mechanism of such change, energy dispersive spectrum (EDX) was used to investigate the compositional change after treatment.

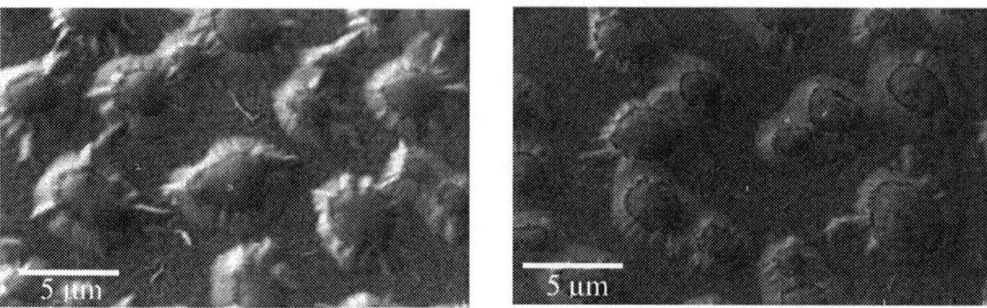

Figure 2. Effect of BOE treatment to morphology change of Ohmic contacts by scanning electron microscopy.

Elemental mapping was performed to study the element distribution of Ohmic contacts by SEM and EDX. Figure 3 shows the element mapping of Au, Ga, Al, Ti, and Ni. The ring region was composed of Au while the island is composed of Al and Ni. The field area consisted of Ga mainly. Ti, on the other hand, was uniformly distributed among the whole field and can't conclude from this resolution.

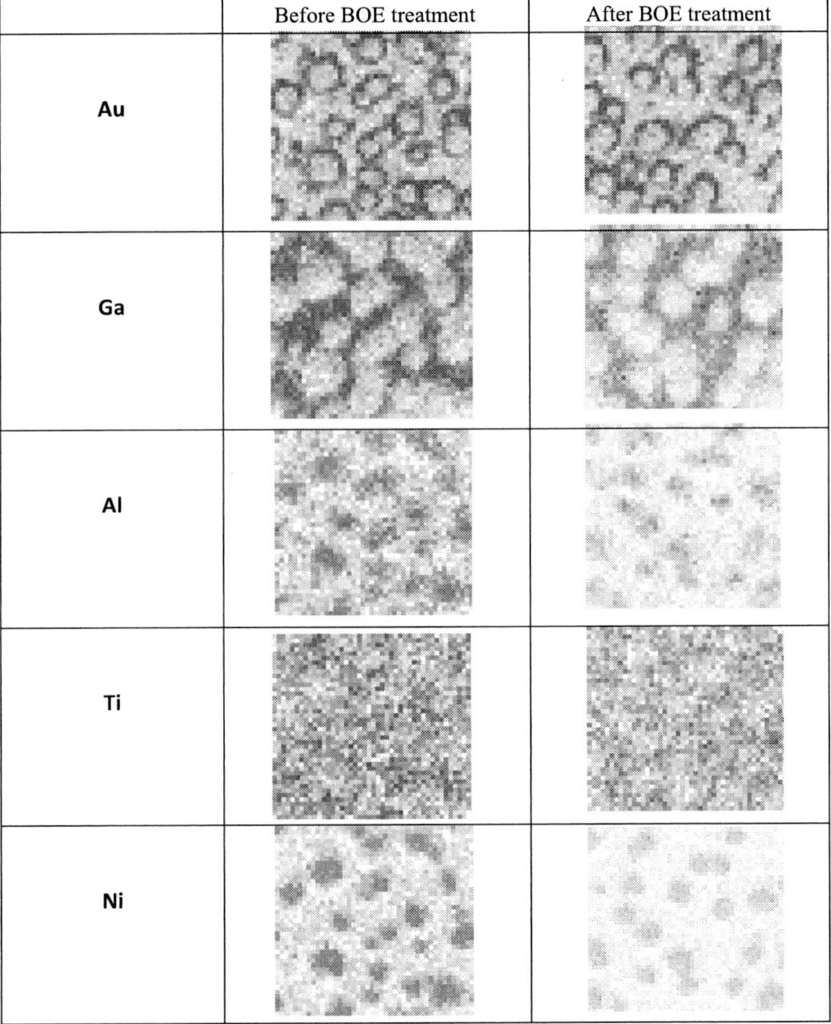

Figure 3. Elemental mapping of Ohmic contacts before and after treatment by energy dispersive spectrum (EDX).

Quantitative analysis was also performed in the EDX analysis and shown in Table 1. As shown in Table 1, Au decreased by 7% although it supposed to be not reactive to BOE. The content of Ti decreased by 0.1% and the contents for the rest of elements, Al, Ni and N, increased by 0.2, 1.3 and 0.6%, respectively. For EDX, the X-rays are generated in a region about 2 microns in depth of the sample. The thickness of the Ti/Al/Ni/Au Ohmic metallization was around 0.3 μm, and thus the EDX signals of Ga and N should be dominated by the bulk GaN even with some Ga or N out-diffusing through the alloyed Ohmic metallization. Thus, it is reasonable to assume that the Ga and N contents did not change significantly. The percent increases of Ga and N content after BOE treatment were due to the decrease of the Au content. By comparing the increases of percent content of Al, it was less than the percent increase of Ga or N. Therefore, besides Ti and Au, the content of Al also decreased after BOE treatment. However the increase of Ti/Al/Ni/Au Ohmic metallization contact resistance should be mainly caused by the decrease of Au content in the BOE treated sample. Although Au is inert to BOE, Au could be removed by the etching of Al underneath this layer.

TABLE 1. Summary of the effect of backside voltage

	Before BOE treatment	After BOE treatment
Ti	3.1	3
Al	7.4	7.6
Ni	4	5.3
Au	55.1	47.5
Ga	26.3	31.7
N	4.2	4.8

To investigate the mechanism of the changes of island and ring area, Auger electron scanning was performed. Ar ion was used to mill the sample and the sample was analyzed after each milling. The depth profile before and after BOE treatment for island and ring area are as shown in Figure 4. Figure 4(a) shows the Auger depth profiling of the island area on the untreated sample. The surface region consisted of C, O, Ti and Al, followed with a 40 nm Ti layer and a layer of 200 nm Ni-Al alloy. Ga clearly diffused throughout the Ni-Al alloy and Ti layer. It has previously been reported that a Ni-Al intermixing layer in these types of contacts was formed after annealing to minimize the interfacial energy [8-9]. After BOE treatment, there was no clear effect of BOE treatment on Ni-Al alloy layer. However, Ti becomes less after treatment. This is expected because Ti would react with BOE as previously discussed. Besides, there was a Ti peak appearing at the Ni-Al alloy and GaN interface for the reference sample, which was not observed for the untreated sample. Zhou *et al.* reported that TiN-based contact inclusion (CIs) form at the GaN surface [7]. The diameter and density of CIs were around 100 nm and 2×10^{-7} /cm^2, respectively. The Auger beam spot size used in this study was around 35 nm. Thus, it is possible that the CI region might be missed during the Auger depth profiling for the untreated sample. Figure 4(c) and (d) show Auger depth profile of the ring area for the untreated sample. There was a surface layer containing C, O, Ti and Al, followed with a Ti layer mixed with Au-Al alloy layer, a thicker Au-Al layer, and an interface layer between Au-Al-Ti alloy layer and GaN. After BOE-treatment, not only was the surface Al was removed, but also some Ti and Au-Al alloyed layers were etched off, as shown in Figure 4(b). This is consistent with EDX data which Au decreased around 7%. Although Au is quite stable with acid solution, the etching of surface Al and Al in Au-Al alloy layers might take away Au. The removal of Au on these surface layers increased the Ohmic metallization sheet resistance.

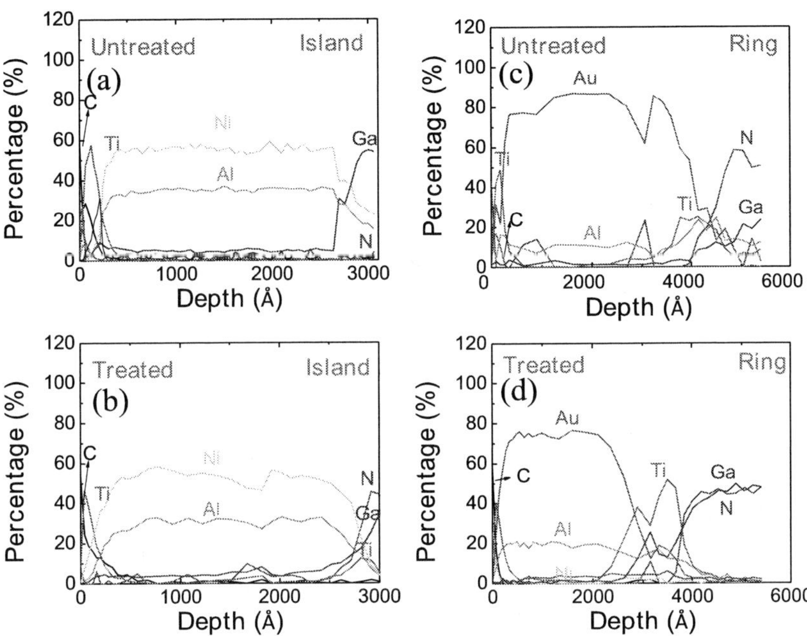

Figure 4. Auger depth profile of (a) untreated island, (b) treated island, (c) untreated ring, and (d) treated ring area.

Conclusions

In summary, the degradation of Ohmic metallization exposed in BOE was studied. The contact resistance of Ohmic metallization increased significantly after treatment in BOE for 2 minutes. Moreover, after annealing, there were island-like structures surrounded by Au-Al alloy rings and a field area between the islands. The BOE etching occurred mainly at the island and ring areas instead of the field area between the islands. The increase of sheet resistance was due to the etching of surface Al and Ti and the loss of Au in the island and ring areas.

Acknowledgments

The work performed at UF is supported by an U.S. DOD HDTRA Grant No. 1-11-1-0020 monitored by Dr. James Reed and a NSF Grant No. ECCS-1445720 monitored by Mahmoud Fallahi. A portion of this research was conducted at the Center for Nanophase Materials Sciences, which is a DOE Office of Science User Facility. Author would like to thank Dr. Pat McKeown in Evans Analytical Group for very fruitful discussion of Auger results.

References

[1] Y.-F. Wu, A. Saxler, M. Moore, R. Smith, S. Sheppard, P. Chavarkar, T. Wisleder, U. Mishra, and P. Parikh, IEEE Electron Dev. Lett. **25**, 117 (2004).

[2] H. Sun, A. Alt, H. Benedickter, and C. Bolognesi, IEEE Electron Dev. Lett. **45**, 376 (2009).

[3] C.-F. Lo, L. Liu, C. Y. Chang, F. Ren, V. Craciun, S. J. Pearton, Y. W. Heo, O. Laboutin, and J. W. Johnson, J. Vac. Sci. Technol. B **29** (2011).

[4] D.Selvanathan, F. M. Mohammed, A. Tesfayesus, I. Adesida, J. Vac. Sci. Technol. B 22, 2409 (2004)

[5] J. Algueperse, P. Mollard, D. Devilliers, M. Chemla, R. Faron, R. Romana, and J. P. Cuer, *Ullmann's Encyclopedia of Industrial Chemistry*, (Wiley, New York, USA, 2005).

[6] Kirt R. Williams, and Richard S. Muller, Journal of Microelectromechanical System, **5** 256 (1996).

[7] Lin Zhou, Jacob H. Leach, Xianfeng Ni, Hadis Morkoc, and David J. Smith, J. of Appl. Phys. **107**, 14508 (2010).

[8] Zhang Yue-Zong, Feng Shi-Wei, Guo Chun-Sheng, Zhang Guang-Chen, Zhuang Si-Xiang, Su Rong, Bai Yun-Xia, and Lu Chang-Zhi, Chin. Phys. Lett., **25**, 4083 (2008).

[9] C.-F. Lo, L. Liu, C. Y. Chang, F. Ren, V. Craciun, S. J. Pearton, Y. W. Heo, O. Laboutin, and J. W. Johnson, J. Vac. Sci. Technol. B **29** (2011).

Chapter 6

Radiation Effects

ECS Transactions, 69 (14) 121-127 (2015)
10.1149/06914.0121ecst ©The Electrochemical Society

Simulating RF Performance of Proton Irradiated AlGaN/GaN High Electron Mobility Transistors (HEMT)s

[a]Shrijit Mukherjee, [a]Erin Patrick, [a]Mark E. Law

[a]Department of Electrical and Computer Engineering
University of Florida, Gainesville, FL, 32611

AlGaN/GaN High Electron Mobility Transistors (HEMTS) subjected to proton irradiation of fluence up to 10^{14} cm^{-2} were simulated to observe the impact on small signal and RF characteristics. While a lot of work has covered dc degradation of HEMTs, very few studies exist which focus exclusively on RF degradation despite their high frequency applications. This work establishes a simulation framework capable of frequency domain device simulation up to 10 GHz. The role of defects generated by proton irradiation will be studied by carrying out low frequency transconductance simulation and current gain frequency sweeps to examine degradation of cut-off frequency. The primary objective will be to understand the role of partially ionized donor traps near the AlGaN/GaN interface and how their bias dependence and capture time constants may significantly impact cut-off frequency and current gain estimations at high frequencies.

Introduction

Due to their relative radiation hardness, High Electron Mobility Transistors (HEMTs) with an AlGaN/GaN heterostructure are of high interest for space applications. Many studies have been performed to quantify the level of degradation due to proton radiation, with the majority being focused on the effects on the DC or steady-state performance. Fewer studies have evaluated degradation of the radio frequency (RF) performance due to proton radiation (1-3). Chen et al. caution that degradation due to proton irradiation has a more measurable impact on the RF performance in comparison to the DC performance (1). In particular, gate lag and increases in both channel resistance and device capacitance due to fast bulk and surface traps contribute more notably to RF degradation than to DC degradation (1). Understanding the role of trap dynamics is key to understanding the exact mechanisms behind degradation.

In previous work, we have found that donor traps have a larger role than previously thought on the magnitude of DC performance degradation in proton-irradiated devices (4). Acceptor traps are always ionized near the conducting channel (two-dimensional electron gas, 2DEG) and the amount of donor compensation determines the extent of degradation. Thus we are most interested in exploring the role of dynamic donor ionization upon RF performance of AlGaN/GaN HEMTs. In this work, we build the simulation framework for frequency-domain small-signal analysis of semiconductor devices capable of accurate simulation up to RF frequencies. Using existing device models we simulate the impact of radiation on RF performance parameters. Primary focus will be on the role of donor traps and how the dynamics of electron capture near the 2DEG can lead to under- or overestimation of device degradation.

Device Structure

Two HEMT structures with different gate lengths were defined. The larger device with a gate length of 1 μm is identical to the one described earlier in a study of degradation of DC performance upon radiation (4). The smaller device has a gate length of 0.2 μm whose overall structure is similar in dimension to the one used in experimental study of RF performance degradation upon proton irradiation (2). Both devices share the same vertical stack as in Fig (1) consisting of substrate of SiC 1um thick, 1.8um of GaN buffer layer, 15 nm AlGaN barrier layer and 0.3um thick metal gate with SiN passivation on either side. Also, the gate for both devices are symmetrically separated from drain and source, with 1.5 um separation for larger device and 1 um for the smaller device. An Al mole fraction of 0.25 was used in AlGaN. P-type doping of 2 x 10^{14} cm^{-3} concentration was added to the GaN buffer to fit well with near threshold DC behavior.

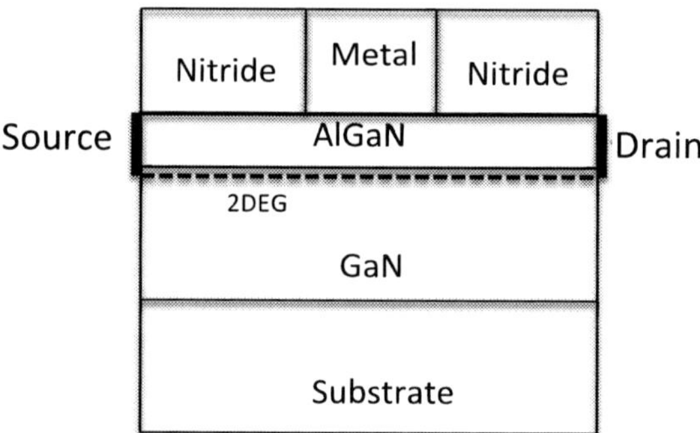

Figure 1. HEMT structure used for simulation. Gate length L_G=1um was used for transconductance simulation and L_G=0.2um for cut-off frequency extraction.

Simulation Methodology

FLOODS (Florida Object Oriented Device Simulator) TCAD is a finite-element technique based partial differential equation solver capable of solving Poisson and continuity equations [1-3] to simulate electrical performance.

$$\nabla^2 \psi = -\frac{q}{\varepsilon}[p - n + Doping] \qquad [1]$$

$$\frac{\partial n}{\partial t} = \frac{1}{q}\nabla . J_n, \qquad \frac{\partial p}{\partial t} = \frac{1}{q}\nabla . J_p \qquad [2]$$

$$J_n = -q\mu_n n\nabla\phi_{fn}, \qquad J_p = q\mu_p p\nabla\phi_{fp} \qquad [3]$$

In Equation 1, ψ is electrostatic potential, q is charge, and ε is material permittivity, n is the density of electrons, and p is the hole density. A doping term is also included. For the continuity [2] and current density equations [3] – with 'n' representing electrons and 'p' representing holes- J_n and J_p are the current densities, μ_n and μ_p are the carrier mobilities and ϕ_n and ϕ_p are the quasi-Fermi levels, which can be related to the electrostatic potential through the Boltzmann relation.

Given a mesh which defines the structure and partial differential equations to define the physics, the solver discretizes the non-linear system and assembles the equations in linear form as $JX = B$, where J is the Jacobian matrix, X is a vector of solution variables which are to be solved for and B is a vector consisting of boundary conditions to be applied to the system. For steady-state device simulation, Newton based iteration techniques are used to increment the solution variable until the correct solution is obtained. In this case, the variables being solved for are electrostatic potential ψ and electron and hole quasi-Fermi levels ϕ_{fn} and ϕ_{fp}.

In order to study the impact of radiation on RF parameters and their degradation, small signal capability was built into FLOODS using a Sinusoidal Steady State Analysis (S3A) technique as described in (5). This technique solves the device equation system in the frequency domain as a steady state perturbed by an infinitesimal signal. The small signal assumption allows linearization of the device around the DC bias point. Consequently the AC system can then be defined in the form: $[J + jD]X = B$ where D contains the time dependent element of the continuity equations. For computational reasons, the AC system takes the form of equation [4], where D is now a diagonal matrix with $\omega = 2\pi f$ as its diagonal elements, and X_R and X_I are real and imaginary components of the solution variables.

$$\begin{bmatrix} J & -D \\ D & J \end{bmatrix} \begin{bmatrix} X_R \\ X_I \end{bmatrix} = \begin{bmatrix} B \\ 0 \end{bmatrix}, \qquad [4]$$

J remains the Jacobian obtained at the DC bias point, while only elements representing contact nodes where the AC supply is applied are set for vector B.

Electrically active defects generated by radiation damage have been modeled as acceptor or donor traps and incorporated in the Poisson equation as:

$$\nabla^2 \psi = -\frac{q}{\varepsilon} [p - n + Doping + Donor - Acceptor] \qquad [5]$$

Acceptor traps, which are Ga vacancies at an energy level of $E_v+1.0$ eV can be considered to be completely ionized and hence considered static. However, donor traps in the form of N vacancies, owing to their energy level close to the conduction band, $E_c-0.1$ eV, particularly those close to the AlGaN/GaN interface, will be sensitive to biasing and hence partially ionized. The ionization function for a donor trap is shown in Equation 4.

$$\frac{N_D^+}{N_D} = \frac{1}{1 + 2e^{\frac{E_F - E_T}{kT}}} = F_D(E_F, E_T), \qquad [4]$$

where N_D is the total donor trap concentration, N_D^+ is the ionized donors, and E_F and E_T are the electron quasi-Fermi level and trap level respectively. Equation 4 was implemented in FLOODS with a slight modification (4). An impulse distribution of the traps at their respective energy level leads to unstable oscillation during iteration process; therefore a Gaussian distribution around the trap level was implemented.

Capture events will dominate as well if speed of operation is slower than capture lifetime. To simulate this scenario, donors have been set up as dynamic in nature by including expressions derived from capture emission statistics, which were then added to the continuity equation:

$$\frac{\partial n}{\partial t} = \frac{1}{q}\nabla.J_n - K_f n N_D^+ + K_r(N_D - N_D^+) \tag{6}$$

$$\frac{\partial N_D^+}{\partial t} = -K_f n N_D^+ + K_r(N_D - N_D^+) \tag{7}$$

where K_f is capture rate dependent on capture lifetime, K_r is emission rate dependent on energy level of the trap. The effect of varying emission and capture rates can then be observed in both transient and frequency domain simulations.

Results and Discussion

<u>Steady State IV and Transconductance</u>

Previous simulation of radiation effects on their DC performance was done with FLOODS (4); however, simulation of transconductance was unable to be performed. Figure 2 illustrates the small signal capability now implemented in FLOODS. The figure

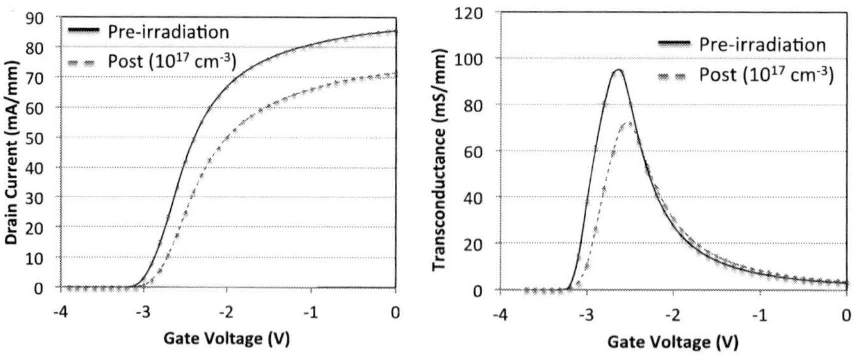

Figure 2. IV characteristic for $V_{DS} = 0.5V$ before and after irradiation (trap concentration of 10^{17} cm^{-3}) (left) and transconductance as a function of gate voltage (right) under same conditions.

shows simulation results of the degradation in transconductance and DC IV characteristics for a HEMT pre- and post radiation. The trap concentration of 10^{17} cm^{-3} corresponds to 5 MeV proton radiation at a fluence of 2×10^{14} cm^{-2} as estimated by TRIM. A decrease in transconductance and positive shift in its peak value is observed similar to experimental data (1, 6).

Figure 3 shows the relation between peak transconductance and vacancy concentration. Despite some positive shift in the peak, degradation can only be observed for trap densities greater than 10^{16} cm^{-3}. This corresponds to fluence values of 10^{13} cm^{-2} and above for a proton energy of 5 MeV. A consistent reduction in transconductance of $16 - 19\%$ was observed over a range of drain voltages from 0.5 to 2.5V when comparing unradiated devices with devices having vacancy densities of 10^{17} cm^{-3}.

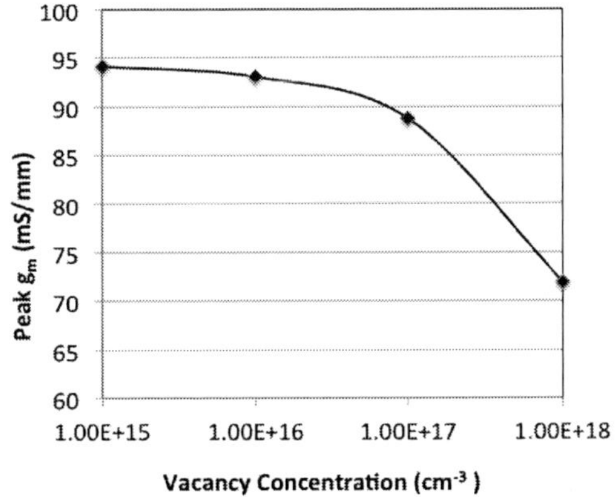

Figure 3. Peak transconductance as function of varying trap concentration for radiated HEMTs at $V_{DS} = 0.5V$

Current Gain Frequency Sweep

For testing the sensitivity of cut-off frequency f_T to radiation, current gain was observed for a frequency-sweep from 1kHz- 10GHz. Experimental measurements of f_T are typically carried out at peak transconductance bias point (3). Frequency sweep simulations were carried out on the smaller devices for comparison with available experimental values (3). All devices were biased at a drain voltage of V_{DS}=0.5V. Pre-radiation device gate was biased at V_{GS}=2.8V, while devices irradiated with a fluence of 10^{14} cm^{-2} was biased at V_{GS}=2.6V, their respective peak transconductance bias points. A radiated device was simulated for 3 different donor capture lifetimes (A) very slow ($\tau=10^{-1}$ sec), (B) within the frequency range window ($\tau=10^{-8}$ sec), and (C) very fast ($\tau=10^{-15}$ sec).

Comparing between pre and post-radiation devices, the reduction in current gain can be observed which will translate to reduction in cut-off frequency (3). For post-radiated devices, the gate bias is low enough to ionize donors close to the electron gas, which allows us to observe the effect of varying capture time constants. Post-radiation device (C) has the lowest gain, as the traps are fast enough to capture, hence reduce drain current. Traps in device (A) on the other hand, are too slow to respond at these frequencies, remain ionized and can be considered static at these frequencies. Device (B) traps are initially fast enough having gain as low as device (C), but beyond 0.1 GHz, they remain ionized and gain coincides with that of device (A). This translates to the trend: devices having the fastest donor traps will have lowest cut-off frequency f_T.

One key factor to be taken into consideration is cut-off frequencies may not be very accurate if obtained directly. The physical models assumed may break down at frequencies higher than 10 GHz. Therefore, intermediate values have been compared, assuming extrapolation of obtained curves will provide correct values for f_T.

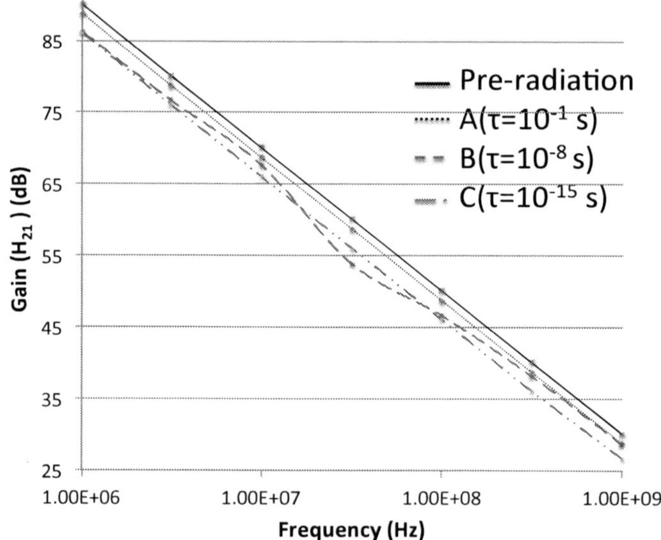

Figure 4. Frequency sweep of current gain H_{21} (I_{GS}/I_{DS}) with 1V AC applied at gate. Radiated devices A, B and C have equal trap concentrations with different donor capture time constants.

Conclusion

Small signal analysis capability was build into FLOODS TCAD solver to study radiation effects on RF performance parameters of AlGaN/GaN HEMTs. Transconductance observed for different device sizes has shown consistent tolerance to trap concentrations up to 10^{16} cm^{-3} for both acceptor and donor traps. For higher level trap concentrations associated with a proton fluence of 10^{14} cm^{-2}, degradation in transconductance such as a positive shift and decrease in peak value have shown good agreement with experimentally obtained results. Current gain and cut-off frequency simulations at RF frequencies up to 10 GHz showed reasonable accuracy for both pre- and post-radiation samples exhibiting expected degradation in both gain and f_T. Behavior of dynamic donor traps also matched theoretically expected results in the frequency domain.

References

1. J. Chen et al., *IEEE Trans. Nucl. Sci*, **61**, 6, 2959 (2014).
2. L. Liu et al., *J. Vac. Sci. Technol.* B **31**, 022201 (2013).
3. B. Luo, et al., *Applied Physics Letters* **79** 14, 2196 (2001).
4. E. Patrick, et al., *ECS J. Solid State Sci. Tech.* **4** 3, Q21 (2015).
5. S. E. Laux, *IEEE Trans. Electron Devices*, **32**, 2028 (1985).
6. A. Kalavagunta, et al., *IEEE Trans. Nucl. Sci*, **55** 4, 2106 (2008).

ECS Transactions, 69 (14) 129-135 (2015)
10.1149/06914.0129ecst ©The Electrochemical Society

Study on effect of proton irradiation energy in AlGaN/GaN metal-oxide semiconductor high electron mobility transistors

Shihyun Ahn[a], Chen Dong[a], Weidi Zhu[a], Byung-Jae Kim[a], Ya-His Hwang[a], Fan Ren[a], Stephen J. Pearton[b], Gwangseok Yang[c], Jihyun Kim[c] and Ivan I. Kravchenko[d]

[a] Department of Chemical Engineering, University of Florida, Gainesville, Florida 32611
[b] Department of Material Science and Engineering, University of Florida, Gainesville, Florida 32611
[c] Department of Chemical and Biological Engineering, Korea University, Seoul 136-713, Korea
[d] Center for Nanophase Materials Sciences, Oak Ridge National Laboratory, Oak Ridge, Tennessee 37830

The effects of proton irradiation energy on Al_2O_3/AlGaN/GaN metal oxide semiconductor high electron mobility transistors (MOSHEMT) were studied. MOSHEMTs were irradiated at different irradiation energies of 5 MeV, 10 MeV, or 15 MeV with a fixed proton dose of 5×10^{15} cm^{-2}. After the 5 MeV, 10 MeV and 15 MeV proton irradiation, MOSHEMTs' saturation current at gate voltage of 1V were reduced by 95.3, 68.3 and 59.8% and maximum transconductance were reduced by 88, 54.4, and 40.7% respectively. The carrier removal rate of the irradiation energy employed in this study was in the range of 127-289 cm^{-1}, having lower energy with the highest removal rate.

Introduction

AlGaN/GaN high-electron mobility transistors (HEMTs) are promising for high power and high frequency applications, including inverter units in hybrid electric vehicles, advanced radar systems, satellite-based communication networks and space communication systems[1-3] their wide energy bandgap, high sheet carrier concentration and high electron mobility. However, the high gate leakage and drain current collapse in these conventional Schottky gate metal devices at high voltage operation limits the stability and performance of the HEMTs[4, 5]. MOSHEMTs have been widely employed to reduce the gate leakage current and passivate surface traps, with Sc_2O_3, MgO, SiO_2 and Si_3N_4[2, 6-10], as the most effective gate dielectrics to solve the HEMTs' reliability issue. Al_2O_3 deposited by atomic layer deposition (ALD) has shown to be the promising gate oxide for its advantages in excellent conformability, low defect density, low stress, and excellent adhesion[5, 11]. Moreover, Al_2O_3 oxide's high dielectric constant (k~10), high breakdown field (5-10 MV/cm), excellent thermal stability, and chemical stability against reaction with AlGaN boost its promising gate dielectric candidacy for MOSHEMT. In this work, the effect of proton irradiation energy on ALD-deposited Al_2O_3 AlGaN/GaN MOSHEMT's was investigated. A fixed dose of 5×10^{15} cm^{-2} proton with irradiation energies of 5 MeV, 10 MeV and 15 MeV were used in the study. The MOSHEMTs' dc characteristics, sheet and contact resistance, threshold voltage were compared before and after proton irradiation.

Experimental

The AlGaN/GaN heterostructure was consisted of metal-organic chemical vapor deposition (MOCVD) grown 5μm GaN buffer, 21 nm of un-doped $Al_{0.24}Ga_{0.76}N$, and a 5nm undoped GaN cap on sapphire substrate. Inductively coupled plasma (ICP) system with Cl_2/Ar plasma was used to achieve the device isolation through mesa-etching. Electron-beam deposition was used to obtain Ti/Al/Ni/Au (20nm/125nm/45nm/100nm) based Ohmic metallization with subsequent rapid thermal annealing at 850°C for 45 sec in a N_2 ambient. A 10 nm Al_2O_3 was deposited by atomic layer deposition at 300°C. Figure 1 shows transmission electron microscopy (TEM) images of Al_2O_3 on calibration silicon wafer. Gate metal, Ni/Au (20nm/80nm) was deposited by electron-beam evaporator. Proton irradiations were conducted at the Korean Institute of Radiological & Medical Sciences (KIRAMS) using a MC 50 (Scanditronix) cyclotron.

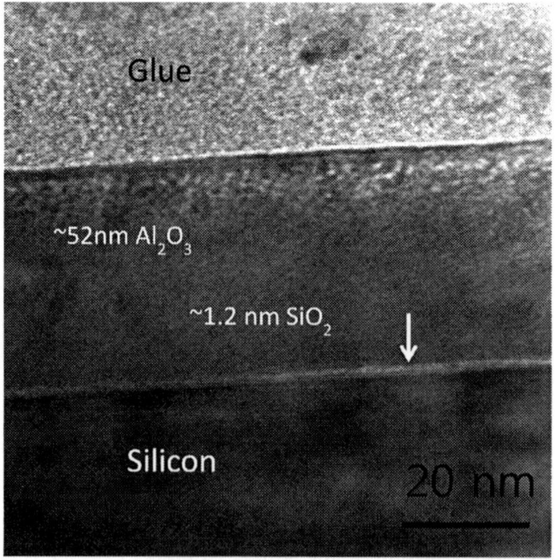

Figure 1. Cross-sectional TEM image of ALD deposited Al_2O_3 on silicon

4156 HP parameter analyzer was employed to measure MOSHEMT dc characteristics. Transmission line measurements (TLM) were used to determine sheet resistance (R_S), transfer resistance (R_T), and contact resistance (R_C) prior and post to proton irradiation. Stopping and range of ions in matter simulations (SRIMs) were used to simulate the proton penetration depth and proton irradiation generated vacancy concentration distribution.

Results and Discussion

Figure 2 shows the SRIM simulation of vacancies created as a function of proton penetration into the Al_2O_3/AlGaN/GaN MOSHEMT structure. The irradiated proton causes atomic displacements and is the main cause of carrier loss due to trapping into these defects in the end-of-range. The energy loss is dominated by electronic stopping which leads to ionization and heating nearer the surface. Thus, the damage induced by proton irradiation is focused near the end-of-range at the penetration depths of 120, 392 and 794 μm for 5, 10 and 15 MeV, respectively. The protons irradiated at 15 MeV protons will penetrate through the wafer since the wafer thickness is around 500 μm. Thus 15MeV irradiated MOSHEMT would get the least amount of damage created in the active regions of the MOSHEMTs. The 2DEG channel is located 36 nm below the Al_2O_3 surface of the MOSHEMT. Therefore, the damage created by the high energy proton in the MOSHEMT structure is uniform but much lower than in the tail region of the damage profile; the simulated vacancy concentrations are 6.6×10^{18}, 3.2×10^{18}, 1.8×10^{18} cm^{-3} for 5, 10 and 15 MeV irradiation, respectively. Table I shows that the sheet resistance, R_S, of the 5 MeV proton-irradiated MOSHEMT was increased more than 10 times and the sheet resistances were increased around 2.5 times for 10 and 15 MeV proton irradiated MOSHEMTs. These experimental measurements are consistent with the SRIM simulation results where 5 MeV simulation showed the most damage in the 2DEG. Besides the increases of the sheet resistance, the MOSHEMT transfer resistance, R_T, and specific sheet resistance, R_C, also increased inversely proportional to the proton energy, as shown in Table I.

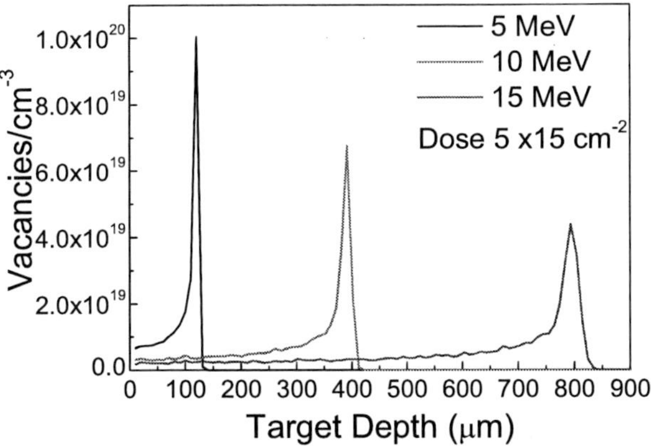

Figure 2. SRIM simulation on vacancies created by proton irradiation as a function of target depth in the AlGaN/GaN MOSHEMT structure

Table I. Effect of proton irradiation on transfer resistance, sheet resistance and specific contact resistance of Al_2O_3/AlGaN/GaN MOSHEMT

	Condition	Rs ($\Omega/\square$)	Rt (Ω-mm)	Rc (Ω-cm^2)
5MeV	Pre-	493	0.48	4.7×10^{-6}
	Post-	5713	6.85	8.2×10^{-5}
10MeV	Pre-	489	0.51	5.3×10^{-6}
	Post-	1254	1.00	7.9×10^{-6}
15MeV	Pre-	497	0.37	2.8×10^{-6}
	Post-	1260	0.76	4.6×10^{-6}

Figure 3(a) shows the drain I-V curves of MOSHEMTs pre- and post- proton irradiation at 15MeV fluence of 5×10^{15}/cm^2.15 MeV irradiated MOSHEMT's saturation drain current at $V_G = 1$V was reduced by 59.8%. A much larger saturation drain current reduction, 95%, was observed for the MOSHEMTs irradiated with proton energy of 5 MeV, and there was only 68.3% drain current reduction for the 10MeV proton irradiation as shown in Figure 3(b). The reduction of saturation drain is due to both decrease in electron density in the channel and in saturated carrier velocity. Displacement damage from the collisions between the incident protons and the semiconductor lattice, defect centers are introduced. Free carriers were captured in these defect centers, reducing carrier density and conductivity of irradiated MOSHEMTs[12-14]. The mobility is affected both by the dielectric/semiconductor interface roughness[15, 16] and the scattering from defect centers created through proton irradiation in the vicinity of the MOSHEMT channel[13, 17]. The decrease of carrier concentration and mobility are indicated in the decrease of initial slope of the drain I-V curves in the linear region[18]. The low field linear region of the drain I-V curves were used to calculate the electron mobility and sheet carrier concentration in the 2DEG channel through charge control model[19]. The extracted electron mobility and sheet carrier concentration show reductions by 63.5 to 99% and 30.3 to 55%, respectively, depending on the proton energy. The electron mobility and sheet carrier concentrations of proton irradiated MOSHEMTs are shown in figure 4(a). Figure 4(b) shows inversely proportional carrier removal rate to the irradiated proton energy, RNC = -16.0•E (MeV) + 362.7, where RNC is carrier removal rate and E is proton energy. The carrier removal rate is the ratio of carrier concentration decrease divided by the fluence of irradiated protons. The lower energy proton irradiation created more non-ionizing energy loss thus more damage in the 2DEG is occurred. Therefore higher removal rate is produced in lower proton irradiation energy. These removal rates are similar to metal gate HEMTs irradiation[20] under similar conditions and show that the presence of the dielectric is not influencing the response of the devices to proton irradiation.

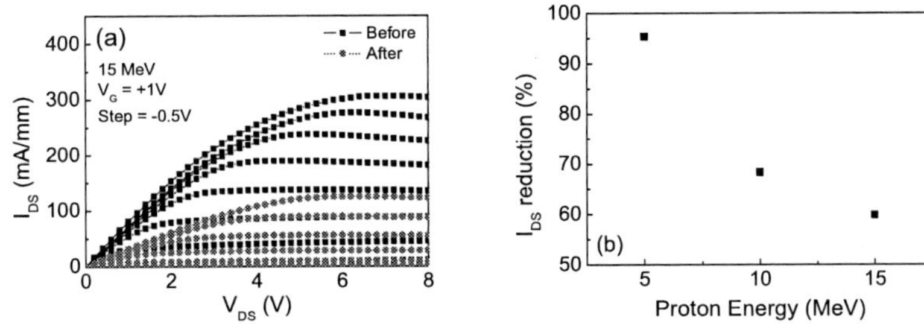

Figure 3. (a) MOSHEMT's drain I-V pre- and post- proton irradiation with fluence of 5 × 10^{15} cm^{-2} at 15 MeV (b) Reduction of saturation current as function of proton irradiation energies at fixed fluence of 5 × 10^{15} cm^{-2}

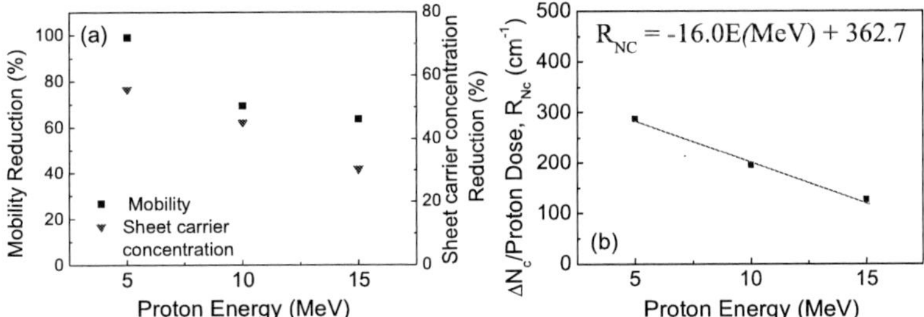

Figure 4. (a) Reduction in mobility and sheet carrier concentration as a function of proton irradiated energies at fixed fluence 5 × 10^{15} cm^{-2} (b) Carrier removal rate of MOSHEMT as a function of proton irradiation energy (cm^{-1})

Figure 5 (a) shows transfer characteristics from the MOSHEMTs pre- and post-irradiation at 10MeV. The extrinsic transconductance, G_M, was reduced by 54.4 % and threshold voltage, V_{TH}, was positively shifted by 1.96V. These degradations were mainly due to reduction in the carrier density and electron mobility which are caused by the displacement damage from ion bombardment. Figure 5 (b) shows more severe degradation of the G_M and a larger positive V_{TH} shift were observed for 5MeV proton irradiation energy, with the G_M decreased around 88% and V_{TH} shifted by 2.89V. Ion penetration was shallower for the lower irradiation energy. Thus, more damage was occurred around the AlGaN/GaN interface.

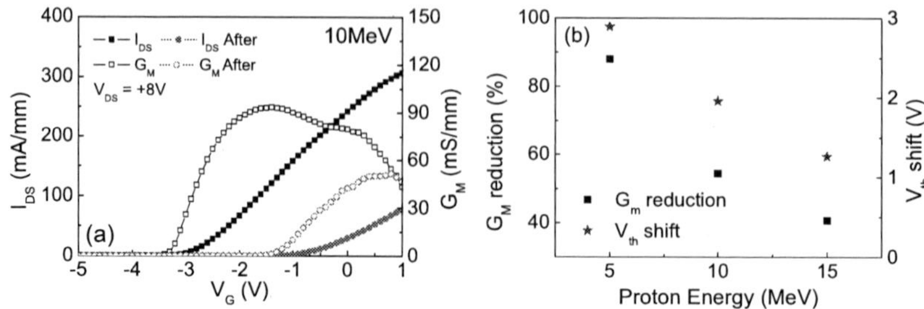

Figure 5. (a) MOSHEMT's transfer characteristics of pre- and post- proton irradiation with fluence of $5 \times 10^{15}\, cm^{-2}$ at 10MeV (b) Reduction of g_m and shift in threshold voltage as function of proton irradiation energies at fixed fluence of $5 \times 10^{15}\, cm^{-2}$

Conclusion

Fluences of $5 \times 10^{15}\, cm^{-2}$ protons at irradiation energies of 5 MeV, 10 MeV, and 15 MeV were used to study the effect of proton irradiation energy on Al_2O_3/AlGaN/GaN MOSHEMTs. All showed significant reduction in carrier mobility, carrier concentration, and degradation in dc characteristics especially for the lowest proton energy irradiation. The severe degradation at lower irradiation energy is due to deep trap generation near the 2DEG channel which increased the resistance of the channel and this defect generation also influenced the positive shift in the threshold voltage.

Acknowledgement

The work performed at UF is supported by an U.S. DOD HDTRA Grant No. 1-11-1-0020 monitored by Dr. James Reed and a NSF Grant No. ECCS-1445720 monitored by Dr. Mahmoud Fallahi. A portion of this research was conducted at the Center for Nanophase Materials Sciences, which is a DOE Office of Science User Facility.

Reference

[1] K. H. Baik, Y. Irokawa, F. Ren, S. J.Pearton, S. S. Park, and S. K. Lee, Solid State Electron. 47, 975 (2003).

[2] M. A. Khan, X. Hu, A. Tarakji, G. Simin, J. Yang, R. Gaska, and M. S. Shur, Appl. Phys. Lett. 77, 1339 (2000).

[3] Y. Irokawa et al., Appl. Phys. Lett. 84, 2919 (2004).

[4] T. Mizutani, Y. Ohno, M. Akita, S. Kishimoto, and K. Maezawa, IEEE T. Electron Dev. 50, 2015 (2003).

[5] P. D. Ye et al., Appl. Phys. Lett. 86, 3501 (2005).

[6] B. Luo, J.W. Johnson, J. Kim, R.M. Mehandru, F. Ren, A.H. Onstine, C.R. Abernathy, S.J. Pearton, A.G. Baca, R.D. Briggs, R.J. Shul, C. Monier and J. Han, Appl. Phys. Lett. 80, 1661 (2002).

[7] J. Kim, B. Gila, R. Mehandru, J.W. Johnson, J.H. Shin, K.P. Lee, B. Luo, A. Onstine, C.R. Abernathy, S.J. Pearton and F. Ren, J. Electrochem. Soc. 149, G482 (2002).

[8] J. Kim, R. Mehandru, B. Luo, F. Ren, B.P. Gila, A.H. Onstine, C.R. Abernathy, S.J. Pearton and Y. Irokaw, Appl. Phys. Lett. 81 (2), 373-375 (2002).

[9] L. Pang, Y. Lian, D.-S. Kim, J.-H. Lee, and K. Kim, IEEE T. Electron Dev. 59, 2650 (2012).

[10] X. Hu et al., Appl. Phys. Lett. 79, 2832 (2001).

[11] Y. Yue, Y. Hao, Q. Feng, J. Zhang, X. Ma, and J. Ni, Sci. China Ser. E. 52, 2762 (2009).

[12] X. W. Hu, A. P. Karmarkar, B. Jun, D. M. Fleetwood, R. D. Schrimpf, R. D. Geil, R. A. Weller, B. D. White, M. Bataiev, L. J. Brillson and U. K. Mishra, IEEE T. Nucl. Sci. 50, 1791 (2003).

[13] H. Y. Kim, J. H. Kim, S. P. Yun, K. R. Kim, T. J. Anderson, F. Ren and S. J. Pearton, J. Electrochem. Soc. 155, H513 (2008).

[14] A. P. Kharmarkar, B. Jun, D. Fleetwood, R. D. Schrimpf, R. A. Weller, B. D. White, L. J. Brillson, U.K. Mishra, IEEE T. Nucl. Sci. 51, 3801 (2004).

[15] R. Oberhuber, G. Zandler and P. Vogl, Appl. Phys. Lett. 73, 818 (1998).

[16] J. Antoszewski, M. Gracey, J. M. Dell, L. Faraone, T. A. Fisher, G. Parish, Y. F. Yu and U. K. Mishra, J. Appl. Phys. 87, 3900 (2000).

[17] B. D. White, M. Bataiev, S. H. Goss, X. Hu, A. Karmarkar, D. M. Fleetwood, R. D. Schrimpf, W. J. Schaff and L. J. Brillson, IEEE Trans. Nucl. Sci. 50, 1934 (2003).

[18] B. Luo, J. W. Johnson, and F. Ren, K. K. Allums, C. R. Abernathy, and S. J. Pearton, R. Dwivedi, T. N. Fogarty, and R. Wilkins, A. M. Dabiran, A. M. Wowchack, C. J. Polley, and P. P. Chow, A. G. Baca, Appl. Phys. Lett. 79, 2196 (2001).

[19] M. Shur, GaAs Devices and Circuits. New York: Plenum, (1986).

[20] L. Liu et al., J. Vac. Sci. Technol. B 31, 022201 (2013) .

ECS Transactions, 69 (14) 137-144 (2015)
10.1149/06914.0137ecst ©The Electrochemical Society

Cathodoluminescence Studies of Gamma-Irradiation Effects on AlGaN/GaN High Electron Mobility Transistors (HEMTs)

A. Yadav[a], M. Antia[a], E. Flitsiyan[a], L. Chernyak[a], I. Lubomirsky[b], and J. Salzman[c]

[a] Department of Physics, University of Central Florida, Orlando, Florida 32816, USA
[b] Department of Materials and Interfaces, Weizmann Institute of Science, Rehovot 76100, Israel
[c] Department of Electrical Engineering, Israel Institute of Technology, Haifa 3200003, Israel

The effects of ^{60}Co gamma-irradiation on AlGaN/GaN High Electron Mobility Transistors (HEMTs) were studied by means of temperature dependent Cathodoluminescence (CL). The CL spectra were examined for several devices with radiation exposures of gamma-ray doses up to 1000 Gy, and for temperature ranging from 25°C to 125°C. Gamma-irradiation causes the CL intensity of HEMTs to decrease by creating defects that act as non-radiative recombination centers. The activation energy was observed to decrease as dosage was incremented to 200 Gy. The decrease in activation energy is related to Compton electron induced increase in carrier lifetime. However for doses above 200 Gy, the activation energy increases as the dose increases. This behavior could be explained by the formation of large electrically active defect complexes through the addition of new radiation defects to the defects formed at low doses.

Introduction

GaN and its related materials have attracted interest for their potential applications in optoelectronic and high-power/temperature electronic devices with relatively low power consumption (1). Because of the unique material properties such as high thermal and mechanical stability as well as high cohesive strength of the N-Ga and N-Al bonds, they are also expected to have substantial resistant to radiation damage and thus should be useful for space and terrestrial environment where radiation is strong. Energetic proton and electron irradiation dominate in studies of GaN radiation resistance compared to a few gamma irradiation studies available, which are generally conducted with high doses of gamma-photons with energy above 1 MeV (2-4). In contrast, an interesting result that has promising practical application is obtained in this study for low doses of gamma radiation.

To better understand the effects of gamma-irradiation on AlGaN/GaN HEMTs, the current state of knowledge regarding the radiation defects created in these materials was taken into account (5-7). There are a number of different types of point defects that can occur in irradiated GaN and AlGaN. These defects include vacancies, interstitials, anti-sites as well as complexes between defects and impurities. The most basic primary defects produced in GaN by radiation are the shallow donor V_N (nitrogen vacancy) with experimentally determined activation energy (ΔE_a) of ~60meV below the conduction band. The double charged donor Ga_i (Gallium interstitial) is deep paramagnetic center with ΔE_a of $\sim 0.8eV$ below the conduction band. The N_i (nitrogen interstitial) is a potential acceptor with ΔE_a of $\sim 1eV$ below the conduction band, which is likely mobile and recombines with existing impurities or vacancy or with defects. The V_{Ga} (gallium vacancy) is the double charge state which produces an acceptor state with ΔE_a of $\sim 1eV$ above the valence band. In AlGaN we can observe many of the same defects, but the energy levels differ due to molecular bonding arrangement that depend on the aluminum mole fraction. The goal of this research is to link defects caused by the radiation to changes in the optical properties of the HEMTs.

Luminescence is a very strong tool for detection and identification of point defects in semiconductors, especially in wide band gap varieties where applications of electrical characterization is limited. In spite of considerable progress made in the last decade in light emitting and electronic devices based on GaN, the understanding and identification of point defects remain surprisingly enigmatic, especially when additional traps are introduced under external irradiation. This can be attributed to a vast number of controversial results in the literature. In this work, the effects of gamma-irradiation on AlGaN/GaN high electron mobility transistors are discussed. The effect of ^{60}Co γ-rays for doses up to 1000 Gy on the luminescence properties of the devices through Cathodoluminescence (CL) measurements are reported, and then a prediction into what defects may be responsible for the radiation effects is suggested.

Experimental

The HEMT devices were grown on Si wafers by Metal Organic Chemical Vapor Deposition (MOCVD) with conventional precursors in a cold-wall, rotating- disc reactor designed from flow dynamic simulations. The growth process was nucleated with an AlN layer to avoid unwanted Ga-Si interactions. The epitaxial stack consisted of an AlGaN transition layer, ~ 800-nm GaN buffer layer, and 16-nm unintentional-doped $Al_{0.26}Ga_{0.74}N$ barrier layer. The nominal growth temperature for the GaN buffer and AlGaN barrier layers was 1030°C.

HEMT fabrication began with Ti/Al/Ni/Au Ohmic metallization and RTA in flowing N_2 at approximately 825 °C. Contact resistance, specific contact resistivity, and specific on-resistance were 0.45 Ω-mm, 5×10^{-6} Ω-cm^2, and 2.2 Ω-mm, respectively. Inter-device isolation was accomplished by using multiple energy N+ implantation to produce significant lattice damage throughout the thickness of the GaN buffer layer. The ion implantation step maintains a planar geometry in the fabricated device and reduces parasitic leakage paths that may exist in passivated, mesa-isolated HFETs. Immediately following implantation, the wafers were passivated with 70-nm-thick SiNx in a PECVD chamber maintained at a base plate temperature of 300 °C. Schottky gate definition was achieved by patterning the gate and selectively removing the SiN_x passivation layer. The contact windows to the Ohmic contact pads were also opened at the same time as defining the Schottky gate. After SiN_x etching, wider gate patterns were redefined with another photolithography step and contact windows to the Ohmic contact pads were also opened. Ni/Au-based gate metallization was deposited on the gate and Ohmic contact pads simultaneously. The wafers were then passivated with another 400-nm layer of PECVD SiN_x at 300 °C. The contact windows were opened by dry etching. There was an additional metal deposition for the HEMT with the source field plate. The field plate was connected to the source terminal and extended by 1 μm out over the gate to the gate-to-drain region. The source to gate distance and channel length of the HEMTs with and without the source field plate were kept constant at 1 and 4.7 μm, respectively.

Devices were exposed to different ^{60}Co gamma-irradiation doses of 100, 200 and 300 and 1000 Gy. Irradiation was performed by NORDION, Inc., at room temperature. During the irradiation, the samples were held in nitrogen ambient, with drain, gate and source contacts kept electrically shorted in order to avoid the influence of self-heating and high electrical field stress effects.

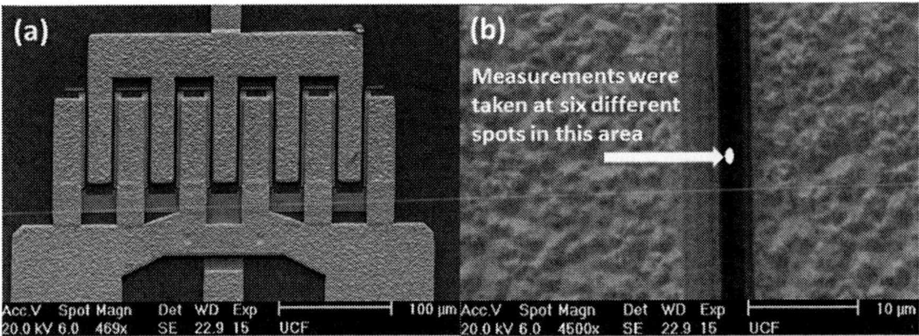

Figure 1. (a) SEM image of AlGaN/GaN device layout. (b) SEM image indicating location of the electron beam spot for CL measurements.

To understand the gamma-irradiation effect on optical properties of AlGaN/GaN HEMTs, CL measurements were performed. CL is the emission of photons of characteristic wavelengths when a material is exposed to high energy electron beam. Extensive scientific information can be obtained by studying the CL spectrum, since its (CL) features depends on the electronic and defect properties of the excited material. CL measurements were obtained on devices prior to gamma-irradiation and then continued on the same devices subjected to various doses of irradiation. Measurements were carried out in a Philips XL 30 Scanning Electron Microscope (SEM) equipped with Gatan monoCL3 system under a 20 kV accelerating voltage. The emitted photons were collected using a single grating (1200 lines/mm, blazed at 500nm) and a Hamamatsu photomultiplier tube (PMT) sensitive to wavelengths ranging from 185-850 nm. The temperature of the sample was varied by 25° C in-situ by the external controller. The SEM electron irradiation was carried out in a spot mode to generate the CL spectrum. In all the measurements, CL spectrums were recorded at different locations in order to avoid the unintentional influence of the electron beam. By comparing the activation energy before and after gamma-irradiation, the defect levels and their dependence on the dose of irradiation was identified.

Results and Discussion

Fig. 2 inset shows the CL spectra of AlGaN/GaN HEMTs taken at different temperatures. It is seen that CL intensities approximately follow an exponential decay with increasing temperature, as expected from the formula (8, 9):

$$I = \frac{I_o}{1 + A\exp^{\left(\frac{-\Delta E_a}{kT}\right)}} \qquad [1]$$

where I_o is the intensity at zero absolute temperature, A is a temperature independent constant, ΔE_a is the activation energy, k is the Boltzmann constant, and T is the temperature in kelvin. The activation energy must represent either a barrier to capture a carrier at non-radiative recombination centers, or the thermal activation energy of such centers. At higher temperatures there is subsequent annihilation of non-equilibrium electron hole pairs generated by the SEM electron beam. It is observed that the frequency of recombination events (CL peak intensity) decreases as the temperature increases since the hole capture cross section is inversely proportional to temperature. CL spectrum also shows characteristic temperature dependence: the emission peak shifts to a higher wavelength (red-shift) as the temperature increases (10, 11, 12). Fig. 2 inset shows the small red shift of approximately 47 meV as temperatures increase from 25°C to 125°C. The slight change in the band gap could be due to the expansion or contraction in the interatomic distance (13).

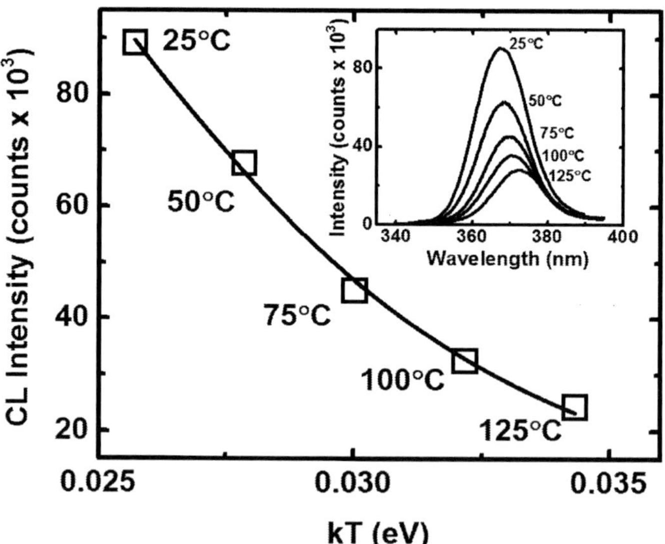

Figure 2. Decay of CL peak intensity with temperature. Inset: CL spectra measurement of AlGaN/GaN reference sample for various temperatures.

Gamma-irradiation produces native point defects and electrically active defect complexes in the material. For a gamma irradiated material, the concentration of a particular defect depends on the flux of radiation; the threshold energy displacement; the thermal stability of the defect species; preexisting defects; and the Fermi level. The decay of CL peak intensity caused by the gamma-irradiation is shown in Fig. 3. The luminescence spectrum has a maximum around 367.52 nm, which corresponds to energy of ~3.37 eV. The luminescence peak intensity decreased by 12.90 % (Sample 1 after 100 Gy) and 37.23% (Sample 2 after 1000 Gy). Radiation causes the decrease in the CL spectrum intensity by creating localized energy levels in the material's band gap. These levels act as a non-radiative recombination centers. There are two different mechanisms responsible for the decrease in the CL intensity for low dose and high dose gamma-irradiation.

The primary defects created in AlGaN/GaN HEMTs by gamma-rays are Frenkel pairs, produced by the Compton electrons having the mean energy of about 600 keV. In this respect, gamma-irradiation, in terms of its impact, is analogical to internal electron irradiation (14). SEM electron beam generates non-equilibrium electron hole pairs, which recombine either via band to band transition or a transition that involves neutral metastable levels. However, if the Compton electron, generated by the gamma-irradiation, gets trapped by the neutral meta-stable levels, recombination cannot proceed. The neutral levels responsible for this phenomenon are most likely associated with nitrogen vacancies (V_N). Trapping the Compton electrons on the V_N-levels prevents radiative recombination of the conduction band electrons (generated by the SEM beam) through these levels. This leads to an increase of the minority carrier life time in the conduction band (15, 16). Since, the intensity of luminescence is inversely proportional to the non-equilibrium carrier lifetime (17), the decay in the CL intensity is observed at low doses (up to ~ 200 Gy).

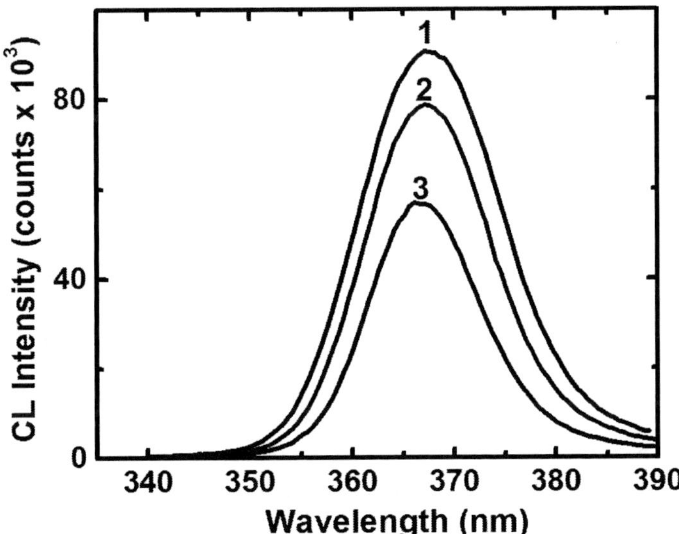

Figure 3. Room temperature CL spectra of AlGaN/GaN devices. Spectrum 1 corresponds to pre-irradiation, Spectrum 2 and 3 corresponds respectively to 100 Gy and 1000 Gy of gamma-irradiation doses.

As the dose of gamma-irradiation increases, additional deep traps due to nitrogen vacancies are introduced in AlGaN/GaN devices. The gamma-introduced deeper defect levels lead to an increased trapping of carriers and dispersion of charge. For higher dose of gamma-irradiation, non-equilibrium carrier scattering on radiation-induced defects dominates over the Compton electron-induced carrier lifetime, and as a result, decrease in the CL intensity is observed.

Cathodoluminescence measurements performed on the devices show that the activation energy decreases for low doses up to 200 Gy and then increases for further higher irradiation as shown in Fig. 4 and Table I. The behavior of the decrease in the activation energy after the low dose gamma-irradiation and the increase for higher doses has been previously studied using Electron Beam Induced Current (EBIC) measurements and reported in (18, 19). EBIC measurement reports also show the increase in diffusion length, L, and decrease in the activation energy up to 200 Gy (Fig. 4 inset). As discussed above, gamma-irradiation causes internal electron irradiation, which in-turn, leads to the trapping of the Compton electrons on the point defect levels. Compton electron trapping on the levels prohibit the recombination pathway. This leads to an increase in the non-equilibrium carrier lifetime (τ) and, in turn, carrier diffusion length L ($L = \sqrt{D\tau}$, where D is carrier diffusivity) (20). Note that carrier diffusivity ($D = \frac{kT}{q}\mu$, where k is Boltzmann constant, T is temperature, μ is mobility and q is elementary charge) is also a temperature and mobility-dependent quantity which can affect the diffusion length and therefore the calculated activation energy. The difference in the activation energy from optical (CL) and electrical (EBIC) measurements could be due to temperature dependence of carrier diffusivity (L is measured as a function of temperature) or due to different V_N levels (or its complexes) in the material's forbidden gap responsible for capturing the Compton electrons. The results of CL and EBIC measurements indicate a decrease in the activation energy up to 200 Gy due to the same underlying mechanism of Compton electron-induced increase in non-equilibrium carrier lifetime. Starting with the dose of 300 Gy, the increase in the activation energy is observed. This is consistent with higher density of deep traps induced by larger dose of gamma-irradiation. The increase in the activation energy for high doses may also indicate the creation of large electrically active defect complexes through the addition of new radiation defects to the defects formed at low doses.

The present work demonstrates the increase in the CL activation energy in response to low dose of gamma-irradiation. The reported result has significant implications for GaN devices since it provide important information about tailoring the semiconductor material properties via controlled irradiation induced defects. It also provides an opportunity for increasing the heterojunction device's performance and reliability, especially for space and terrestrial applications.

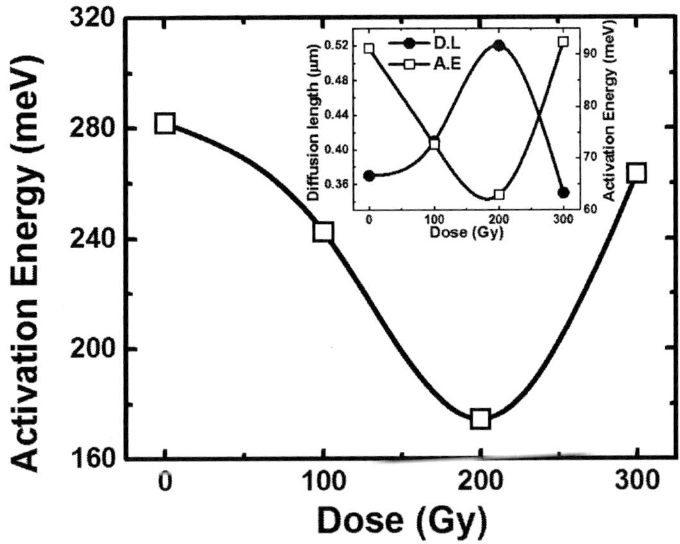

Figure 4. Impact of low dose gamma-irradiation on CL activation energy in AlGaN/GaN HEMTs. Inset: Dependence of experimental diffusion length (closed circles) and associated activation energy (open squares) in AlGaN/GaN HEMTs on irradiation dose from EBIC measurements (**19**).

Table I: Impact of gamma-irradiation on the activation energy (meV) of AlGaN/GaN HEMTs. (*** NI: Not Irradiated)

	0 Gy	100 Gy	200 Gy	300Gy	1000 Gy
Sample 1	300±20	236±14	NI	NI	NI
Sample 2	280±18	NI	NI	NI	380±10
Sample 3	330±22	238±16	173±15	263±11	NI
Sample 4	280±25	252±22	180±12	262±13	NI
Sample 5	320±16	236±12	170±18	265±10	NI
Average	282±36	240±6.7	174±4.1	263±1.2	380

Acknowledgement

This research is supported by NATO SfP project# 984662.

References

1. R. Quay, *Gallium Nitride Electronics*, Springer Series in Material Science, (2008).
2. A. Polyakov, S. J. Pearton, P. Frenzer, F. Ren, L. Liu and J. Kim, *J. Mat. Chem. C*, **1**, 877, (2013).
3. T. B. Elliot, *Trends in Semiconductor*, Nova Science Publishers Inc., p.135, New York (2005).
4. S. J. Pearton, R. Deist, R. Ren, L. Liu, A. Y. Polyakov and J. Kim, J. Vac. Sci. Technol. A 31, 5, 050801 (2013).
5. T. L. Tansley and R. J. Egan, *Phys. Rev. B*, **45**, 19, 10942 (1991).
6. D. C. Look, D.C. Reynolds, J. W. Hemsky, J. R. Sizelove, R. L. Jones and R. J. Molnar, *Phys. Rev. Lett.*, **79**, 2273 (1997).
7. M. A. Reshchikov and H. Morkoc, *J. App. Phys.*, **97**, 061301 (2005).
8. Y. Lin, E. Flitsiyan, L. Chernyak, T. Malinauskas, R. Aleksiejunas, K. Jarasiunas, W. Lim, S. J. Pearton and K. Gartsman, Appl. Phys. Lett., 95, 092101 (2009).
9. D. S. Jiang, H. Jung and K. Ploog, *J. Appl. Phys.*, **64**, 1371 (1988).
10. C. Prall, M. Ruebesam, C. Weber, M. Reufer and D. Rueter, *J. Cryst. Growth*, **397**, 24 (2014).
11. K. P. O'Donnell and X. Chen, *Appl. Phys. Lett.*, **58**, 25, 2924(1991).
12. B. Monemar, *Phys. Rev. B*, **10**,676 (1974).
13. A. Saha, N. Manna, *Optoelectronics and Optical Communications*, University Science Press, 5 (2011).
14. V. V. Emtsev, V. Y. Davydov, V. V. Kozlovskii, V. V. Lundin, D. S. Poloskin, A. N. Smirnov, N. M. Shmidt, A. S. Usikov, J. Aderhold, H. Klausing, D. Mistele, T. Rotter, J. stemmer, O. Semchinova and J. Graul, *Semicond. Sci. Technol.*, **15**, 73 (2000).
15. L. Chernyak, W. Burdett, M. Klimov and A. Osinsky, *Appl. Phys. Lett.*, **82**, 3680 (2003).
16. L. Chernyak and W. Burdett, *MRS Proceedings*, 764 (2003).
17. O. Lopatiuk, A. Osinsky, L. Chernyak, *Proceedings of SPIE*, 6121:H1, (2006).
18. A. Yadav, E. Flitsiyan, L. Chernyak, Y.H. Hwang, Y.L. Hsieh, L. Lei, F. Ren, S. J. Pearton, and I. Lubomirsky, *Rad. Eff. Def. Sol.*, **170**, 2 (2015).
19. A. Yadav, E. Flitsiyan, L. Chernyak, Y. H. Hwang, Y. L. Hsieh, L. Lei, F. Ren, S. J. Pearton, and I. Lubomirsky, *MRS Proceedings* 1792 (2015).
20. K. Kumakura, T. Makimoto, N. Kobayashi, T. Hashizume, T. Fukui, and H. Hasegawa, *Appl. Phys. Lett.*, 86, 052105 (2005).

Chapter 7

Group IV Materials and Devices

ECS Transactions, 69 (14) 147-156 (2015)
10.1149/06914.0147ecst ©The Electrochemical Society

Enhanced Performance Designs of Group-IV Light Emitting Diodes for Mid IR Photonic Applications

J. D. Gallagher[a], C. L. Senaratne[b], C. Xu[a], P. M. Wallace[b], J. Menéndez[a], and J. Kouvetakis[b]

[a] Department of Physics, Arizona State University, Tempe, AZ 85287-1504, USA
[b] Department of Chemistry and Biochemistry, Arizona State University, Tempe, AZ 85287-1604, USA

This paper describes $Ge_{1-y}Sn_y$ (y = 0.0-0.11) light emitting diodes (LEDs) with a broad range of compositions extending from pure Ge to Sn concentrations where the alloy becomes a direct gap material (y = 0.09) and beyond. The devices are grown on Si(100) platforms using ultra-low temperature depositions of highly reactive Ge and Sn hydrides. The device fabrication adopts two new photodiode designs. The first employs n-Ge/i-$Ge_{1-y}Sn_y$/p-$Ge_{1-y'}Sn_{y'}$ hetero-structure stacks possessing a single defected interface between the n-Ge and i-$Ge_{1-y}Sn_y$ layers. The second consists of a homo-structure n-$Ge_{1-y}Sn_y$/i-$Ge_{1-y}Sn_y$/p-$Ge_{1-y}Sn_y$ design. The devices show low dark currents and the electroluminescence (EL) spectra show concomitant redshifts of the peak emission position and intensity increase near direct gap Sn compositions. The EL of the homo-structure design displays a large efficiency enhancement relative to the hetero-structure devices. The fabrication of ternary $Ge_{1-x-y}Si_xSn_y$ diodes with ~9% Sn and 3-9% Si contents is also discussed.

Introduction

Quasi-direct gap group-IV materials offer the potential to enhance modern photonic capabilities by incorporating lasers and light emitting diodes (LEDs) directly on Si-based logic circuits due to the small 140 meV separation between the conduction band Γ and L-valleys in Ge. Initial success in this field was achieved by growing heavily doped n-type Ge using delta doping techniques and creating biaxial tensile strain in layers grown on Si(100) by exploiting the mismatch in thermal expansion coefficients between the two materials through high temperature annealing processes (1,2). The tensile strain states created in these materials reduced the separation between the direct (E_0) and indirect gap (E_{ind}) signals leading to nearly direct gap Ge that showed photoluminescence (PL) and electroluminescence (EL) (3-5). This research culminated in reports of an optically pumped waveguide laser and an electrically pumped Ge laser on Si thereafter (6,7). However the Ge on Si system has a limited range of wavelength tunablility, a constraint that is imposed by CMOS thermal budgets dictating the range of tensile strain states that can be achieved.

An alternative approach to direct gap group-IV materials with extended IR capabilities and tunable band gaps is to alloy Ge with Sn (8). There have been a number of computational and experimental studies demonstrating that incorporating small fractions of Sn into the Ge parent lattice reduces the separation between the E_0 and E_{ind} band gaps and redshifts them in energy relative to Ge, analogous to the effect of biaxial tensile strain on pure Ge (9-11). Most notably, several PL studies have reported a decrease in separation between the conduction band valleys coupled with a redshift of the E_0 and E_{ind} signals leading to an indirect to direct gap crossover at ~9% Sn (12-15). Efforts to mature $Ge_{1-y}Sn_y$ materials technology into operational devices has ushered in reports of various hetero-structure diode devices incorporated onto Si in *pin* geometry. These show a redshift of both the absorption edges and EL peaks as a function of increasing Sn content for devices with *i*-layer Sn compositions above and below the threshold of direct gap materials (16-20). In conjunction with these developments made on $Ge_{1-y}Sn_y$ photodiodes, an optically pumped waveguide laser operating at low temperatures has also been demonstrated (21), a promising sign that the report of an electrically injected $Ge_{1-y}Sn_y$ laser may be in the not so distant future.

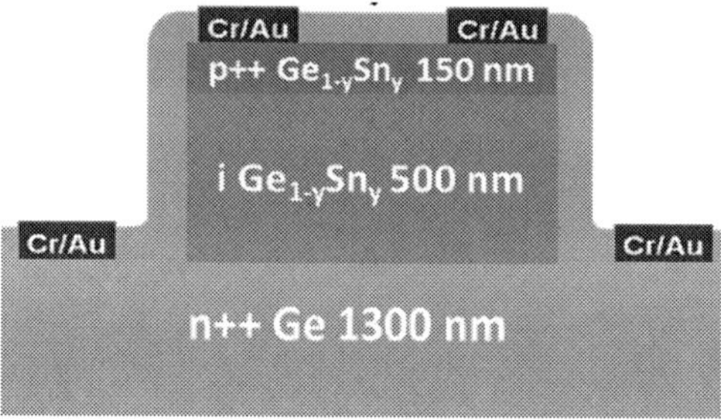

Figure 1. Schematic illustration of the basic hetero-structure (n-Ge/i-Ge$_{1-y}$Sn$_y$/p-Ge$_{1-y}$Sn$_y$) diode grown upon Si. These devices contain a single defected interface thereby reducing the deleterious effects of mismatch-induced dislocations. The ubiquitous Ge platform that is used as a bottom contact is easily produced and readily doped using well known *in situ* methodologies.

To this end this study pursued systematic fabrication and characterization of new types of *pin* photodiodes with compositions of 0-11% Sn. Devices are built following the design depicted in Figure 1. They comprise of a thick Ge bottom contact layer doped *n*-type, an active $Ge_{1-y}Sn_y$ region, and a *p*-type $Ge_{1-y}Sn_y$ top electrode. The latter is grown lattice matched to the intrinsic region creating i/p junctions devoid of interfacial defects that act as recombination traps. These devices feature only one defected interface between the i-$Ge_{1-y}Sn_y$ and n-Ge layers, thus this construction represents an improved design with higher efficiencies compared to initial Ge/i-$Ge_{1-y}Sn_y$/Ge prototypes that contain two defective interfaces. In addition a more advanced design was also explored containing exclusively $Ge_{1-y}Sn_y$ *pin* stacks which are completely devoid of defected interfaces in contrast to the device depicted in Figure 1. This architecture exhibits higher emission efficiencies as described in the subsequent sections, indicating that defects represent the main barrier to reaching optimal functionality in this class of materials. Finally, the degree of directness is also explored in high-Sn content (~9%) ternary $Ge_{1-x-y}Si_xSn_y$ diodes at varying Si concentrations up to 9% to assess the potential of this material system as a thermally robust alternative to $Ge_{1-y}Sn_y$-based optoelectronic devices.

Materials Growth, Characterization, and Device Fabrication

<u>Sample Growth and Characterization</u>

Growth of the $Ge_{1-y}Sn_y$ devices begins with the deposition of a thick (~1 μm), low defect density, n-Ge buffer layer on Si(100) with activated donor P concentrations in the vicinity of 2×10^{19} cm^{-3}, as discussed in prior work (22-24). Next the wafers are loaded into a chemical vapor deposition (CVD) reactor to grow the i-$Ge_{1-y}Sn_y$ material. These growths are done following well-described literature methods through gas-source reactions of Ge_3H_8 and SnD_4 at low temperatures (25,26), yielding thick (300-700 nm) single phase alloys with Sn compositions up to 11% that are largely strain-relaxed on the Ge buffer. For the growth of ternary i-$Ge_{1-x-y}Si_xSn_y$ materials, the gas-source molecule Si_4H_{10} is added to the reaction mixtures (27,28). The device stacks are completed with the growth of a p-$Ge_{1-y'}Sn_{y'}$ top electrode. This is achieved through CVD reactions of Ge_2H_6, SnD_4, and B_2H_6. The top contact layers range in thickness from 100-200 nm with 1.0-10×10^{19} cm^{-3} activated acceptors. The Sn content (y') of these layers is carefully chosen to be $y' < y$ to enhance the overall quantum efficiency of the device, but with the difference $y - y'$ small enough to ensure that the p-layer grows pseudomorphic to the active i-layer. This architecture represents a significant design advantage by eliminating the delirious effects of carrier scattering by defects at the top i/p interface. However these devices are not defect free since strain relaxation occurs at the bottom n-Ge/i-$Ge_{1-y}Sn_y$ interface. In an attempt to improve the emission performance by eliminating the defects generated by strain relaxation at the bottom interface, we modified the hetero-structure design to create a lattice matched homo-structure analog in which all pin constituent layers contain approximately equal amounts of Sn. This approach delivered strain relaxed and lattice matched junctions among the device components. A prototype sample was produced by growing a 6-7% Sn n-type contact layer with thickness 500 nm on a Ge buffered platform using $P(SiH_3)_3$ as the source of P atoms. This was followed with the growths of a 7% Sn i-layer and a 6% Sn p-layer as described above.

Following growth of the completed device stacks, all samples were characterized by Rutherford backscattering spectrometry (RBS), X-ray diffraction (XRD), and spectroscopic ellipsometry to measure the various layer thicknesses, compositions, crystallinity, strain states, and doping levels (20). Secondary ion mass spectroscopy was employed to corroborate the activated doping densities measured by ellipsometry and to verify uniform doping profiles with sharp interfaces across the device stack. STEM-corrected atomic resolution images and XTEM micrographs were acquired on a JEOL ARM 200F operated at 200 kV and a JEOL 4000 EX at 400 kV, respectively (19,20). Analysis of the top i/p interface showed no defects in the field of view and lattice matching in the plane of growth, confirming the pseudomorphic formation of the top p-layer on the underlying i-layer. This notion is supported by XRD measurements that showed matching of the in-plane lattice constants between these two layers (19,20). In the hetero-structure devices the analysis of the bottom n/i interface primarily showed short stacking faults and 60° dislocations that penetrated a short distance into the n-Ge contact layer, leaving the active i-layer generally devoid of threading defects (20).

Additionally, dark field images taken with STEM showed homogeneous colors on both sides of the interface. Since these images are Z-contrast sensitive it substantiates the notion of nearly perfect random Sn substitution with no precipitation or phase segregation in the Ge host, despite the significant fraction approaching 11% in the alloy material.

Device Fabrication

The diodes are fabricated similar to protocols previously described in Ref. 29. Briefly, a protective ~150 nm SiO_2 film is deposited on the as-grown epitaxial sample stack. The mesa patterns are then defined using a positive photoresist exposure and development procedure. Next the protective oxide is removed between the mesas to expose the epilayer material for reactive ion etching using BCl_3 plasma. The etch time is carefully chosen so that the sample is etched at least 200 but no more than 600 nm into the bottom n-type contact layer. A passivating SiO_2 film is then deposited with thickness optimized to minimize this layer's reflectance to light with energy near the band gap of the active material in the device. A second photolithography process is performed in image-reversal mode to define the metal contact patterns. Following this, the passivating oxide is cleared out in the contact regions and the remaining photoresist is used as a mask for thermal evaporation of 20/200 nm Cr/Au metal stacks on the sample. The device is finalized by liftoff of the photoresist and excess metal in acetone followed by an organic descum in oxygen plasma.

Device Electrical, Optical Characterization, and Results

Hetero-structure GeSn Electrical and Optical Characterizations

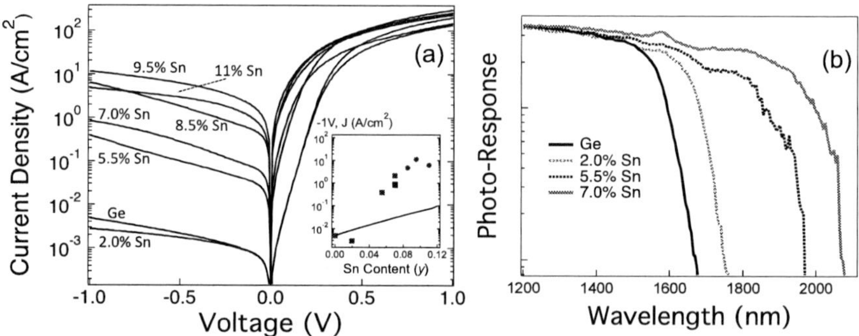

Figure 2. (a) Dark current characteristics of hetero-structure $Ge_{1-y}Sn_{1-y}$ devices across the composition range presented. Inset plots the -1V current levels (points) relative to a theoretical model (solid line) that assumes no defects in the devices. (b) Optoelectronic response at 0V bias on samples with active layers up to 7% Sn demonstrating an extension of the absorption edge beyond 2 μm.

The completed devices are characterized by current-voltage (*IV*) and spectral responsivity measurements to elucidate their electrical and optical properties as a function of Sn content. Figure 2(a) compares *IV* curves for a series of *pin* samples across the 0-11% Sn concentrations showing rectifying diode behavior in all cases. The reverse bias current for the 2% Sn device is lower than that of the pure Ge sample in spite of the significantly larger active layer thickness in the latter sample (530 nm vs. 800 nm, respectively). This result indicates that the electrical properties are independent of the presence of the Ge-Sn bonds or lattice distortions arising from local bonding configurations in the alloy. In contrast to the $Ge_{0.98}Sn_{0.02}$ device, the dark current densities of the more concentrated samples increase by three orders of magnitude from 3 mA/cm^2 to 6 A/cm^2 for the $Ge_{0.89}Sn_{0.11}$ device. In the 5-7% Sn devices a surge in dark current density is observed. This is less pronounced in the 8.5-11% Sn samples where the currents appear to converge within a narrow range spanning 5-10 A/cm^2 indicating that this window may represent a saturation limit. In order to better understand the contribution to the dark current generated by the presence of defects at the n-Ge/i-Ge$_{1-y}$Sn$_y$ interface, we plot the -1V current values as points vs Sn content in the inset of panel (a). The data are compared with a theoretical model (solid line) which plots the expected compositional dependence of the dark current density at -1V bias assuming only Shockley-Read-Hall (SRH) recombination and ignoring defect-based current generation mechanisms. According to SRH theory, the reverse bias current goes as the inverse exponential of the band gap such that lower band gap materials produce larger reverse bias currents, as shown by the line which increases by almost 2 orders of magnitude over $y = 0.00$-0.11. Comparing the SRH line with the points, we note that samples with i-layer concentrations $y > 0.05$ are at least an order of magnitude greater than expected based on SRH theory alone. Thus the difference between the points and the line represents the component that defects, likely formed by strain-relaxation at the n-Ge/i-Ge$_{1-y}$Sn$_y$ interface, contribute to the electrical properties in these devices.

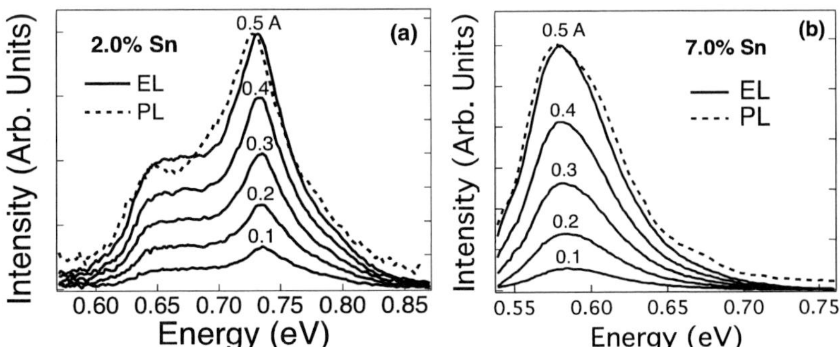

Figure 3. Current-intensity and EL/PL relationships for (a) 2.0% and (b) 7.0% Sn devices. The EL intensities increase super-linearly as a function of injected current in both cases and the peak positions between EL and PL experiments show excellent agreement.

Figure 2(b) plots the normalized optical responsivity absorption edges for hetero-structure *pin* Ge$_{1-y}$Sn$_y$ devices ranging from 0-7% Sn. These experiments are performed at room temperature under 0V bias. The plots show a systematic redshift of the absorption edge to longer wavelengths coinciding with successively higher Sn concentrations incorporated into the active region. These results support the notion that alloying Ge with Sn reduces the fundamental band gaps in the alloy $Ge_{1-y}Sn_y$ system relative to Ge, and demonstrates the applicability of these materials for integrated long wavelength detection systems.

EL experiments were carried out at room temperature as described in prior work using extended InGaAs and PbS detectors (20). Figure 3(a) shows a series of EL spectra taken as a function of current from a 530 nm thick $Ge_{0.98}Sn_{0.02}$ device depicted as solid black lines. The spectrum at 0.5 A is compared with the photoluminescence counterpart measured from an equivalent single layer reference grown on Ge buffered Si using the same methods. The main peak is attributed to E_0 emission and the low-energy shoulder is assigned to E_{ind} transitions. It is apparent that the relative intensity and position of the direct/indirect signals are very similar for both the PL and EL. Figure 3(b) shows the injection current dependence of the EL spectra for a $Ge_{0.93}Sn_{0.07}$ device. In this case the peak energy is significantly redshifted relative to the 2.0% emission energies and the spectra show only a single peak from the E_0 transition. This supports the notion that $Ge_{1-y}Sn_y$ alloys smoothly cross over to direct gap materials along the 0-10% Sn composition range since the E_{ind} signal is obscured at Sn concentrations greater than 7%. The PL spectrum of the same sample was collected from an as-deposited film prior to processing into mesa structures and is overlaid with the 0.5 A EL counterpart showing that the line shapes of the PL and EL emission peaks are closely matched. For both devices the EL intensity increases super-linearly as a function of injection current in the range of 0.1 to 0.5 A while the peak maximum position remains unchanged. This behavior has been previously reported for Ge and $Ge_{0.98}Sn_{0.02}$ diodes that were grown directly on Si (30).

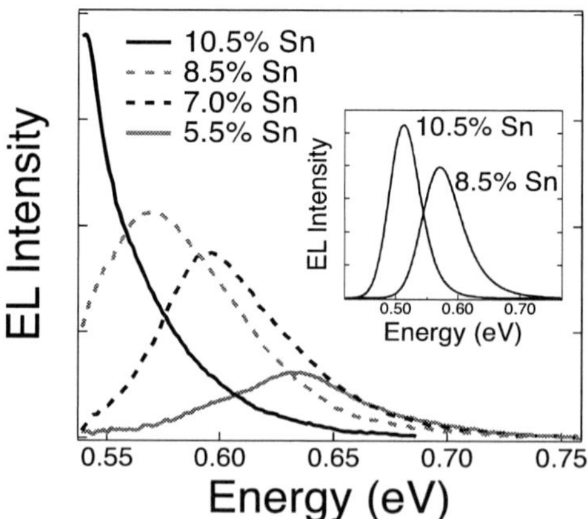

Figure 4. Room temperature EL spectra of hetero-structure $Ge_{1-y}Sn_y$ diodes with similar active layer thicknesses. The plots show a redshift of the E_0 and E_{ind} signals, a merging of these signals, and an increase in peak intensity as the Sn content is systematically increased. Inset compares the EMGs obtained from the 8.5% and 10.5% Sn samples EL spectra collected using the PbS detector.

The EL spectra of samples with Sn contents spanning 5.5-10.5% and *i*-layer thicknesses approximately 400 nm are plotted in Figure 4. The spectra are collected at 0.5 A and show strong peaks corresponding to E_0 emission. The peak maxima shift to lower energies and the emission intensities increase with Sn content, as expected. A sharp rise in signal amplification is observed as the alloys transition to direct gap materials near ~9% Sn. In addition to the main peak, the spectrum of the 5.5% Sn device also shows a weak lower energy shoulder assigned to the E_{ind} transition. This feature coalesces with the main E_0 peak for $y \geq 0.07$ due to the reduction in separation between the direct (Γ) and indirect (L) conduction band edges with increasing Sn content. The sharp cutoff observed for the 10.5% Sn spectrum is due to the InGaAs detector limit. To obtain the full peak profile for this material we used a PbS detector with an extended range down to 2700 nm. The results are shown in Figure 4 inset, which compares exponentially modified Gaussian (EMG) fits to the EL data acquired using this detector for the 8.5% and 10.5% Sn devices. The emission peaks illustrate the expected shift in band gap energy and intensity enhancement vs composition. These EMG fits to the spectral data have been routinely applied in prior work to measure E_0 gaps with meV accuracy for alloys grown on Ge buffered Si (13).

<u>Homo-structure GeSn Optical Characterization</u>

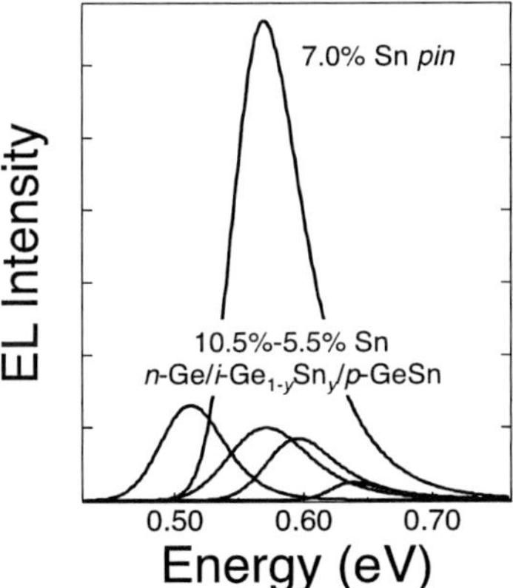

Figure 5. Plots of the EMG fits to the 0.5 A EL data for various GeSn LEDs. The lower-intensity cluster of samples contain a defective *n*-Ge/*i*-Ge$_{1-y}$Sn$_y$ interface. The high-intensity peak is from a lattice-matched *pin* stack where all active device component layers are ~Ge$_{0.93}$Sn$_{0.07}$ alloys.

To optimize the emission performance of the hetero-structure *pin* devices, we modified the design to create a lattice matched homo-structure analog in which all constituent layers contain approximately equal amounts of Sn. This approach yielded strain relaxed and lattice matched junctions between the contact and active layers of the device, circumventing non-radiative recombination pathways produced by defects at the *n*-Ge/*i*-Ge$_{1-y}$Sn$_y$ interface in the hetero-structure samples. Figure 5 compares the EL intensity from a 7% Sn *pin* homo-structure with the intensities from the 5.5-10.5% Sn samples discussed in Figure 4. The homo-structure device displays a 5-fold enhancement of the EL signal compared to the 7% Sn hetero-structure analog. A noteworthy feature of the homo-structure spectrum is that the peak appears slightly red-shifted relative to its analog, a sign that the long minority carrier diffusion lengths in these Ge-like alloys cause a significant fraction of the radiative recombination to take place in the n-GeSn contact layer. This indicates that the emission intensity enhancement observed here can be explained as a combination of two sources: the elimination of defects at all interfaces, and the excess carriers in the conduction band of the n-type contact layer assisting with radiative recombination pathways. This result is important for the design of future high-efficiency group-IV LEDs and diode lasers operating at long wavelengths, which should be based on tuned $Ge_{1-y}Sn_y$ hetero-structures with tailored alloy compositions, thicknesses, doping densities, and strains aimed at maximizing light emission efficiencies.

SiGeSn Diodes

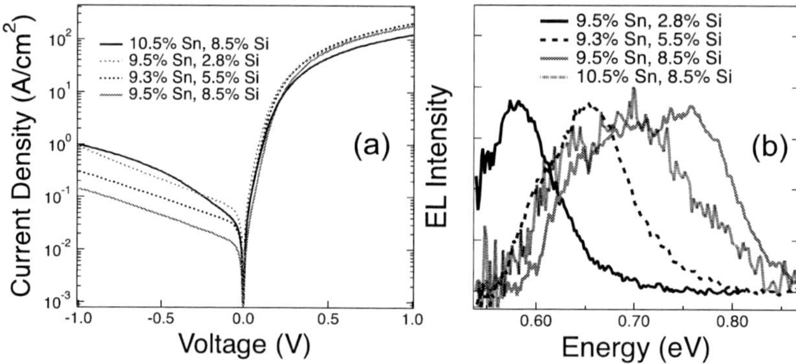

Figure 6. (a) Room temperature IV plots of the full ternary alloy range discussed. These devices show strong rectifying characteristics. (b) EL from $Ge_{1-x-y}Si_xSn_y$ diodes with fixed ~9% Sn contents and Si concentrations in the 3-9% range showing the separation of the E_0 and E_{ind} signals as a function of Si content.

Ternary $Ge_{1-x-y}Si_xSn_y$ alloys represent a burgeoning class of materials afforded by the ability to independently tune the band gap at a fixed lattice constant. The aim here is to investigate regions of the composition space with Sn contents close to where $Ge_{1-y}Sn_y$ becomes a direct-gap semiconductor (9% Sn and above), while adjusting the Si composition to find optimized materials with greater separation between the L and Γ points in the conduction band. Such materials may give target ternary alloys an electronic structure that is more direct than analogous $Ge_{1-y}Sn_y$ materials. Figure 6(a) overlays plots of the IV measurements for a variety of devices representing the full composition range of alloys investigated in this study (~9% Sn and 3-9% Si). Good rectifying properties are demonstrated in the plots over the ±1V bias range shown with ideality factors between 1.10 and 1.20. These devices show dark current levels comparable to or less than that of the $Ge_{1-y}Sn_y$ devices discussed in Figure 2(a) at the same Sn concentrations. The variation in -1V bias dark current densities spans an order-of-magnitude range from ~0.1-1.0 A/cm^2. This suggests that with improved processing conditions, lower dark currents should be achievable in ternary devices relative to analogous $Ge_{1-y}Sn_y$ diodes. This would be due to the incorporation of Si providing closer lattice matching with the underlying n-Ge contact, mitigating some of the mismatch-induced defects. Figure 6(b) plots the normalized room temperature EL from the devices shown in panel (a). The plots show that the incorporation of successively higher Si contents shifts the fundamental band gaps to higher energies and increases the separation between the E_0 and E_{ind} peaks. The band gaps of these materials were determined using modeling procedures discussed in Ref 13. Using this modeling procedure, it was found that the difference $E_0 - E_{ind}$ between the 9.5% Sn, 8.5% Si sample is 110 meV and this value in 10.5% Sn, 8.5% Si device is 70 meV. This indicates that direct-gap ternary alloys should be attainable despite high Si concentrations by incorporating Sn contents that further reduce the quantity $E_0 - E_{ind}$.

Conclusion

In summary we have demonstrated group-IV optoelectronic devices integrated onto Si platforms with optimized designs. These include the development of rectifying hetero-structure $Ge_{1-y}Sn_y$ *pin* diodes with active layer compositions that span beyond the 9% Sn content where this binary alloy system becomes a direct gap material. We also show the first generation of a homo-structure device that shows a significant improvement in light emission relative to the hetero-structure analogs. Using the light emission from the ternary $Ge_{1-x-y}Si_xSn_y$ devices we find that direct gap materials should be attainable even at Si contents up to 10%. These advances are made possible by low temperature reactions and *in situ* doping strategies afforded by the rationally designed hydride precursors employed in the growth stages of the device fabrication, enabling the formation of finely tuned and optimized device architectures.

Acknowledgements

This work was supported by the AFOSR DOD FA9550-12-0208. We gratefully acknowledge the use of the Ira A. Fulton Center for Solid State Electronics Research at Arizona State University.

References

1. R. E. Camacho-Aguilera, Y. Cai, J. T. Bessette, L. C. Kimerling, and J. Michel, *Opt. Mat. Express*, **2** (11), 1462 (2012).
2. X. Sun, J. Liu, L. C. Kimerling, and J. Michel, *Appl. Phys. Lett.*, **95**, 011911 (2009).
3. G. Shambat, S.-L. Cheng, J. Lu, Y. Nishi, and J. Vuckovic, *Appl. Phys. Lett.*, **97**(24), 241102 (2010).
4. S.-L. Cheng, J. Lu, G. Shambat, H.-Y. Yu, K. Saraswat, J. Vuckovic, and Y. Nishi, *Opt. Express*, **17**(12), 10019–10024 (2009).
5. X. Sun, J. Liu, L. C. Kimerling, and J. Michel, *Opt. Lett.*, **34**(8), 1198–1200 (2009).
6. J. Liu, X. Sun, R. E. Camacho-Aguilera, L. C. Kimerling, and J. Michel, *Opt. Lett.*, **35**(5), 697-681 (2010).

7. R. E. Camacho-Aguilera, Y. Cai, N. Patel, J. T. Bessette, M. Romagnoli, L. C. Kimerling, and J. Michel, *Opt. Express*, **20**(10), 11316 (2012).
8. G. He and H. A. Atwater, *Phys. Rev. Lett.*, **79**, 1937 (1997).
9. D. W. Jenkins and J. D. Dow, *Phys. Rev. B*, **36**, 7994 (1987).
10. K. A. Mader, A. Baldereschi, and H. von Kanel, *Solid State Commun.*, **69**, 1123 (1989).
11. M. Bauer, J. Taraci, J. Tolle, A. V. G. Chizmeshya, S. Zollner, D. J. Smith, J. Menéndez, C. Hu, and J. Kouvetakis, *Appl. Phys. Lett.*, **81**, 2992 (2002).
12. J. Mathews, R. T. Beeler, J. Tolle, C. Xu, R. Roucka, J. Kouvetakis, and J. Menéndez, *Appl. Phys. Lett.*,**97**, 221912 (2010).
13. L. Jiang, J. D. Gallagher, C. L. Senaratne, T. Aoki, J. Mathews, J. Kouvetakis, and J. Menéndez, *Semicond. Sci. Technol.*, **29**, 115028 (2014).
14. S. A. Ghetmiri, *et. al.*, *Appl. Phys. Lett.*, **105**, 151109 (2014).
15. J. D. Gallagher, C. L. Senaratne, J. Kouvetakis, and J. Menéndez, *Appl. Phys. Lett.*, **105**, 142102 (2014).
16. H. H. Tseng, K. Y. Wu, H. Li, V. Mashanov, H. H. Cheng G. Sun, and R. A. Soref, *Appl. Phys. Lett.*, **102**, 182106 (2013).
17. M. Oehme *et al.*, *IEEE Photonics Tech. Lett.*, **26**, 2 (2014).
18. W. Du *et al.*, *Appl. Phys. Lett.*, **104**, 241110 (2014).
19. J. D. Gallagher, C. L. Senaratne, P. Sims, T. Aoki, J. Menéndez, and J. Kouvetakis, *Appl. Phys. Lett.*, **106**, 091103 (2015).
20. J. D. Gallagher, C. L. Senaratne, C. Xu, P. Sims, T. Aoki, D. J. Smith, J. Menéndez, and J. Kouvetakis, *Journal of Appl. Phys.*, **117**, 245704 (2015).
21. S. Wirths, *et al.*, *Nature Phontonics*, **9**, 88 (2015).
22. L. Jiang, C. Xu, J. D. Gallagher, R. Favaro, T. Aoki, J. Menéndez, and J. Kouvetakis, *Chem. of Mat.*, **26**(8), 2522-2531 (2014).
23. C. Xu, R. T. Beeler, L. Jiang, G. Grzybowski, A. V. G. Chizmeshya, J. Menéndez, and J. Kovuetakis, *Semicond. Sci. Tech.*, **28**, 105001 (2013).
24. C. Xu, J. D. Gallagher, P. Sims, D. J. Smith, J. Menéndez, and J. Kovuetakis, *Semicond. Sci. Tech.*, **30**, 045007 (2015).
25. C. L. Senaratne, J. D. Gallagher, L. Jiang, T. Aoki, D. J. Smith, J. Menéndez, and J. Kouvetakis, *J. Appl. Phys.*, **116**, 133509 (2014).
26. C. L. Senaratne, J. D. Gallagher, T. Aoki, J. Kouvetakis, and J. Menéndez, *Chem. Mat.*, **26**(20), 6033 (2014).
27. C. Xu, R. T. Beeler, L. Jiang, J. D. Gallagher, R. Favaro, J. Menéndez, and J. Kovuetakis, *Thin Solid Films*, **55**, 177-182 (2014).
28. C. Xu, L. Jiang, J. Kouvetakis, and J. Menéndez, *Appl. Phys. Lett.*, **103**, 072111 (2013).
29. R. Roucka, J. Mathews, C. Weng, R. T. Beeler, J. Tolle, J. Menéndez, and J. Kovuetakis, *J. Quantum Electronics*, **47**(2), 213 (2011).
30. R. Roucka, J. Mathews, R. T. Beeler, J. Tolle, J. Kouvetakis, and J. Menéndez, *Appl. Phys. Lett.*, **98**, 061109 (2011).

ECS Transactions, **69** (14) 157-164 (2015)
10.1149/06914.0157ecst ©The Electrochemical Society

Doping of Direct Gap Ge$_{1-y}$Sn$_y$ Alloys to Attain Electroluminescence and Enhanced Photoluminescence

C. L. Senaratne[a], J. D. Gallagher[b], C. Xu[b], P. E. Sims[a], J. Menéndez[b], and J. Kouvetakis[a]

[a] Department of Chemistry and Biochemistry, Arizona State University, Tempe, AZ 85287-1604
[b] Department of Physics, Arizona State University, Tempe, AZ 85287-1504

Low temperature (T < 300°C) *in-situ* doping protocols were developed for Ge$_{1-y}$Sn$_y$ alloys with compositions above the indirect-to-direct gap crossover point, which were deposited using CVD. Trisilylphosphine (P(SiH$_3$)$_3$) was used as the *n*-type dopant source while diborane (B$_2$H$_6$) was used for *p*-type doping. This enabled the fabrication of *pin* diode structures with compositions up to $y = 0.137$. These photodiodes exhibit electroluminescence at wavelengths up to 2700 nm. Furthermore, it was demonstrated that *n*-type doping enhances the photoluminescence obtained from these materials.

Introduction

Germanium-tin (Ge$_{1-y}$Sn$_y$) alloys have received much attention in recent years due to potential applications in microelectronics and optoelectronics. They have been investigated as stressor materials for use in metal-oxide-semiconductor field-effect transistors (MOSFETs) which make use of strained Ge as a high carrier mobility channel material (1,2). In addition, it has been shown that enhanced mobility can be achieved by employing the alloys themselves as a channel material (1,3,4). Both avenues present strategies that can be used to keep pace with Moore's law by increasing the speed and efficiency of microelectronic devices. From an optoelectronic perspective, Ge$_{1-y}$Sn$_y$ is of interest because it is a group-IV semiconducting binary alloy that has a direct band gap at compositions $y > 0.09$ (5,6). Therefore it provides a pathway for developing efficient optical devices that can be directly integrated with existing Si technologies.

For both applications it is necessary to have procedures that allow *p*- and *n*-type dopant incorporation at low temperatures. This requirement arises from the metastable nature of Ge$_{1-y}$Sn$_y$ alloys with $y > 0.01$ (7). Therefore, low temperature, far-from-equilibrium conditions must be used in the fabrication of alloys with useful Sn contents. Such procedures have the added benefit of being compatible with CMOS processing protocols. Both *p*- and *n*-channel MOSFETs have been produced using MBE and CVD techniques (1,3). For optical applications, the recent trend has been the development of *n*-Ge/*i*-Ge$_{1-y}$Sn$_y$/*p*-Ge or *n*-Ge/*p*-Ge$_{1-y}$Sn$_y$ heterostructure devices (8–11). MBE and CVD techniques have successfully been employed for fabrication of photodiodes based on Ge$_{1-y}$Sn$_y$ using these designs. Electroluminescence covering a wide range of wavelengths up to 2275 nm has been reported from these devices (10–12).

Previous work from our group has demonstrated *in-situ* doping agents for both Ge and $Ge_{1-y}Sn_y$ which can be used in ultra-low temperature CVD depositions (13,14). These precursors have the general formula $M(GeH_3)_3$, where 'M' denotes the group V doping atom, typically P or As. Recently we have extended the options available for low temperature doping by showing that the analogous $M(SiH_3)_3$ (M=P, As) compounds can also be used for this purpose (6). In the same work, optical properties of these materials were elucidated using photoluminescence (PL) studies. In other work, we were able to demonstrate $Si/n\text{-}Ge/i\text{-}Ge_{1-y}Sn_y/p\text{-}Ge_{1-y'}Sn_{y'}$ ($y` < y$) heterostructure devices with active layer compositions of $y = 0\text{-}0.11$, which exhibit emission out to 2600 nm (15,16). In these devices, the *n*-type Ge bottom contact layer was grown using gas source molecular epitaxy (GSME) techniques and doped using $P(GeH_3)_3$ (17). The *p*-type top contact layer was doped using the traditional doping agent B_2H_6, albeit using extremely low temperatures down to 320°C.

In this work we summarize recent progress made in using these doping precursors for synthesis of light emitting devices that have active regions with Sn contents beyond the indirect to direct gap crossover composition ($y > 0.09$), which extend the emission capabilities beyond the mid-IR region.

Growth of intrinsic and *n*-type doped $Ge_{1-y}Sn_y$ alloys

Both intrinsic and *n*-type doped $Ge_{1-y}Sn_y$ alloys investigated in this study were deposited on Ge buffered Si substrates. These substrates were prepared in a GSME reactor, by depositing Ge on Si(100) wafers using Ge_4H_{10} as the source material. The layers were grown to thicknesses of around 1 μm, and then subjected to an *in-situ* anneal which reduces the number of threading dislocations and improves crystallinity. When *n*-type doped Ge buffers were required, the doping was achieved by adding 2% $P(GeH_3)_3$ (in relation to the amount of Ge_4H_{10}) to the mixture. This allowed active carrier concentrations of approximately 2×10^{19} cm^{-3} to be achieved, sufficient for the Ge layer to act as a bottom contact of a *pin* device.

These substrates were cleaved into 45 mm × 45 mm segments and cleaned by dipping in aqueous HF prior to the growth of the $Ge_{1-y}Sn_y$ alloys. The growths were conducted in a hot wall UHV-CVD reactor heated using a single zone furnace. For the growth of intrinsic $Ge_{1-y}Sn_y$ alloys, Ge_3H_8 and SnD_4 were used as precursors, with both compounds being mixed in 3 L bulbs to make a single uniform mixture, then being diluted by research grade H_2. For the intrinsic $Ge_{1-y}Sn_y$ epilayers of composition $y = 0.110\text{-}0.137$ described here, the atomic ratios between Sn and Ge in the mixtures ranged from 0.09-0.13. The concentration of Ge_3H_8 with respect to H_2 was kept constant at around 1% in all mixtures, with the amount of Ge_3H_8 used in a typical growth being ~30 L Torr. Furthermore, a separate mixture of 1.6% Ge_3H_8 in H_2 was used to regenerate the Ge surface *in-situ* prior to the deposition of the $Ge_{1-y}Sn_y$. This refreshing layer deposition was conducted at 340°C for ~25 min. Thereafter continuous flow of Ge_3H_8 was maintained while the chamber cooled to the low temperatures required for $Ge_{1-y}Sn_y$ deposition, in order to maintain a clean surface. This procedure yields refreshing layers ~150 nm thick.

When growing the $Ge_{1-y}Sn_y$ layers, the requirement of low temperatures and the associated low growth rates proved a major challenge for obtaining thick (> 300 nm) epilayers required for superior optical performance. Therefore a two-stage temperature programmed approach was used for the growth of the films described here, in contrast to the single growth temperature used in our earlier work. The $Ge_{1-y}Sn_y$ layer deposition was initiated at a temperature known to be conducive for the growth of $Ge_{1-y}Sn_y$ layers of the target composition. For epilayers in the $y = 0.110$-0.137 range described here, this corresponds to initiation temperatures from 280°C-270°C. Growth was allowed to proceed at this temperature for sufficient time to allow the deposition of a film ~100nm thick. Then the temperature was slowly increased by 5°C-10°C. This procedure is expected to create a strain-relaxed layer of $Ge_{1-y}Sn_y$ at the target composition, which provides a more facile template on which further growth can proceed. The growth was then allowed to complete at this higher temperature. By following this procedure, it was possible to obtain films of $Ge_{1-y}Sn_y$ in the target composition range that were 300-500 nm thick, whilst preventing Sn surface segregation that accompanies $Ge_{1-y}Sn_y$ growth at high temperatures. Figure 1 shows high resolution X-ray diffraction (HRXRD) θ/2θ scan of the (004) reflection and (224) reciprocal space map (RSM) obtained from a 475 nm thick $Ge_{1-y}Sn_y$ layer with $y = 0.13$ grown using the above technique. The Sn content was determined from both Rutherford backscattering spectrometry (RBS) and HRXRD measurements, and are in agreement with each other. The symmetry of the XRD peaks suggest that the change in temperature during growth did not result in any compositional inhomogeneity, while the average growth rate was increased.

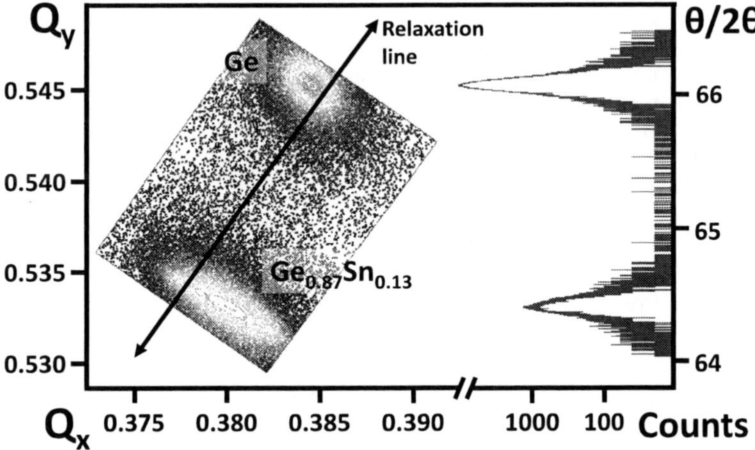

Figure 1 - (224) RSM and (004) θ/2θ scan of a $Ge_{0.87}Sn_{0.13}$ sample with 475 nm thickness grown on Ge buffered Si. Growth was conducted in the temperature range 270°C-280°C.

The same growth procedure was used to obtain films of n-type doped $Ge_{1-y}Sn_y$ films in the composition range $y = 0.09$-0.13. Intrinsic Ge buffers were used as substrates for these experiments. The only modification required was the addition of the doping agent to the precursor mixture. The compound chosen for this purpose was $P(SiH_3)_3$, which is an alternative for $P(GeH_3)_3$. It has a higher vapor pressure and is stable at room temperature, and is therefore easier to use than the germyl compound. We have shown that using the current reactor configuration, only dopant levels of Si are incorporated into the $Ge_{1-y}Sn_y$ lattice with $P(SiH_3)_3$ (6). This is not expected to affect the emission properties of the n-doped alloys described here. By mixing 0.40-0.14 L Torr of $P(SiH_3)_3$ to $Ge_3H_8/SnD_4/H_2$ precursor mixtures, dopant levels of 5×10^{18} cm^{-3} $-$ 2.5×10^{19} cm^{-3} could be attained. The activated dopant levels were determined in this case by IR spectroscopic ellipsometry. The amount of total P incorporated into the lattice was determined using secondary ion mass spectrometry (SIMS), and the two levels are in agreement within error. This indicates that the incorporated P is fully activated, despite the low temperatures used in the growth process.

By growing p-type doped top contact layers, the intrinsic $Ge_{1-y}Sn_y$ samples described above were fabricated into *pin* stacks.

Growth of p-type doped $Ge_{1-y}Sn_y$ alloys

For the *pin* device structures, the compositions of the top layers were selected such that a heterojunction was created between the i- and p-layers, where the top p-layer has lower Sn content than the active i-layer. This allows carrier confinement in the active layer, and also prevents reabsorption of the light emitted from the active region (15). The depositions were carried out in a separate hot-wall CVD reactor which was heated using a three zone furnace, and therefore had a temperature profile suitable for pre-activating the doping precursor B_2H_6. This was supplied via a mass flow controller (MFC) separate from the one used for the flow of the Ge_2H_6 and SnD_4 precursor mixture which deposits the $Ge_{1-y}Sn_y$ of the required composition. The separation of the two mixtures prevents reaction between the different compounds prior to entry into the growth chamber. The p-type layers grown as the top contacts using this method have thicknesses of 100-300 nm and dopant concentrations in the 1×10^{19} cm^{-3} - 9×10^{19} cm^{-3} range. For these epilayers, the compositions were between $y = 0.08$-0.12.

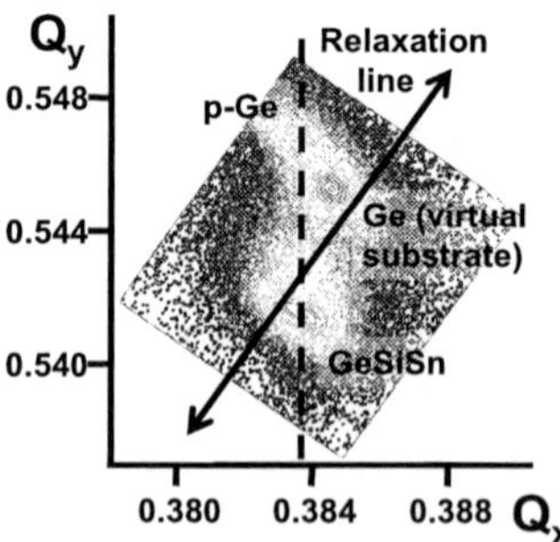

Figure 2 – (224) RSM of the p-type doped Ge layer from grown on a GeSiSn buffer layer. It exhibits a tensile strain of 0.396%. The dashed line illustrates the vertical alignment of the peaks, which is evidence that the Ge layer is pseudomorphic to the GeSiSn substrate.

As with the n-type and intrinsic $Ge_{1-y}Sn_y$ alloys described earlier, deposition of p-type $Ge_{1-y}Sn_y$ alloys near indirect to direct gap crossover composition require low temperatures in the range of 280°C-295°C. In previous work we have used B_2H_6 as the doping agent at temperatures as low as 320°C to dope $Ge_{1-y}Sn_y$ alloys with Sn contents up to $y = 0.05$, which have activated dopant levels as high as 5×10^{19} cm^{-3}. Compared to these earlier samples, a B_2H_6 flux which is approximately two times greater is required for the samples grown at 280°C in order to get the same activated dopant levels. To investigate whether the high fluxes and low growth temperatures result in the incorporation of non-activated B in the $Ge_{1-y}Sn_y$ lattice, these samples were subjected to SIMS analysis using an O_2^+ primary beam to gauge the absolute B content. As an example, the quantification of absolute B in a sample for which the p-type layer was grown at 285°C, gives a concentration of 2.0×10^{19} cm^{-3}, which is in good agreement with the 1.8×10^{19} cm^{-3} value for activated carriers obtained from IR spectroscopic ellipsometry. This indicates that while B incorporation is less efficient at low temperatures, the B within the film is still fully activated.

From the results obtained thus far, it was evident that B incorporation with a high degree of activation could be achieved using our reactor configuration. This led to the investigation of p-type doping of Ge to levels useful in the fabrication of pMOSFETs. Figure 2 shows the (224) RSM of a p-type Ge layer grown on a 71% relaxed $Ge_{0.94}Si_{0.01}Sn_{0.05}$ buffer layer, which in turn was grown on a virtual Ge substrate. The Ge epilayer was grown using Ge_2H_6, with a small amount of SnD_4 added to the precursor mixture to increase the growth rate. The temperature used was 345°C. The resultant film was 110 nm thick, and was found to have an active carrier concentration of 1.0×10^{20} cm^{-3}. SIMS analysis of this sample gives an absolute B concentration of 1.0×10^{20} cm^{-3}. From this data we surmise that despite the very high levels of B present in the epilayer, it can be considered to be fully activated. A property of particular interest in this p-Ge epilayer is that its growth is pseudomorphic to the GeSiSn substrate. This results in a high tensile strain of 0.396%. Such highly tensile strained, p-type doped Ge has applications in the development of Ge based pMOSFETs. The results presented here suggest the viability of our CVD growth techniques for obtaining suitable materials for this purpose.

Device microstructure, electroluminescence and photoluminescence

The pin devices fabricated in this work bear many features in common with similar devices reported earlier spanning Sn compositions from $y = 0.05$-0.11 (15,16). The active $Ge_{1-y}Sn_y$ layer grows relaxed with respect to the virtual Ge substrate. The strain relaxation is accompanied by the formation of defects at the $Ge/Ge_{1-y}Sn_y$ interface. In contrast, the p-type top contact layer grows pseudomorphic to the active $Ge_{1-y}Sn_y$ layer. Therefore no defect formation is observed at the i-p interface. These features are shown in the cross sectional scanning transmission electron microscopy images (XSTEM) given in Figure 3. Figure 3 a) is a medium angle annular dark field (MAADF) image showing the entire device stack. Figures 3 b) and 3 c) are high resolution bright field (BF) images of the top i-p interface and the bottom n-i interface. Typical 60° dislocations arising due to strain relaxation can be seen at the bottom interface, while the defect free nature of the top interface is also evident.

These strain features are clearly illustrated in Figure 4 a), which shows a (224) RSM obtained for the same device stack. A schematic of the completed device obtained after subjecting this stack to standard microelectronics processing procedures is shown in Figure 4 b).

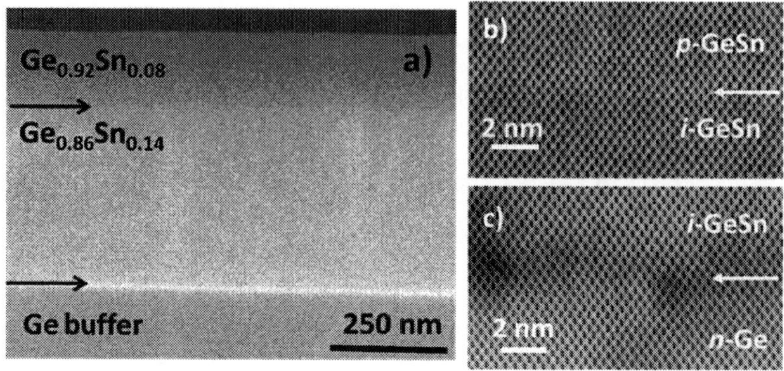

Figure 3 – a) XSTEM MAADF image of a *pin* heterostructure device comprised of an *n*-Ge bottom contact, *i*-Ge$_{0.863}$Sn$_{0.137}$ active layer and *p*-Ge$_{0.92}$Sn$_{0.08}$ top contact. The contrast difference between the layers agrees with the difference in Sn composition, and uniform contrast within the layers indicates compositional homogeneity. b) High resolution BF image of the *p*-GeSn/*i*-GeSn interface. The absence of defects at this heterojunction is attributed to the pseudomorphic growth of the top contact on the active layer. This can be contrasted with the *n*-Ge/*i*-GeSn high resolution image shown in c), which shows 60° dislocations typically observed when GeSn relaxes with respect to the Ge substrate. The interfaces are marked by arrows.

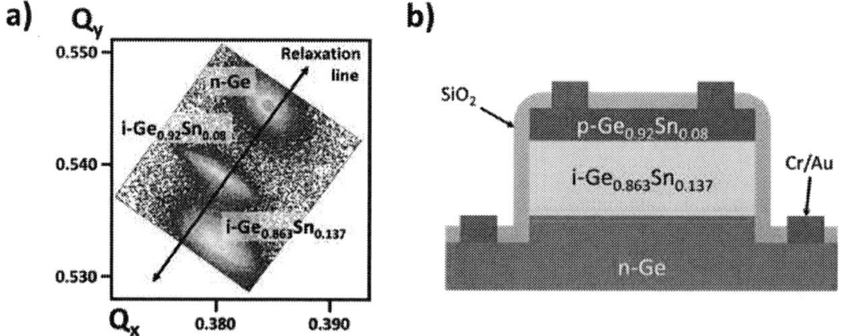

Figure 4 – a) (224) RSM of a *pin* stack composed of a Ge$_{0.863}$Sn$_{0.137}$ active (*i*) layer and a Ge$_{0.92}$Sn$_{0.08}$ *p*-layer. b) Schematic of processed *pin* diode, illustrating contact formation to the top and bottom (*p* and *n*) electrodes.

In Figure 5 a), an exponentially modified Gaussian (EMG) fit to the EL spectrum obtained from a *pin* device with a $y = 0.12$ active region is compared with that obtained from a $y = 0.105$ device reported previously (15). The peak maxima fall at 2400 nm for the latter, and 2600 nm for the former. It can be seen from these results that electrically induced light emission from direct-gap $Ge_{1-y}Sn_y$ alloys is achievable at wavelengths well beyond the mid-IR region.

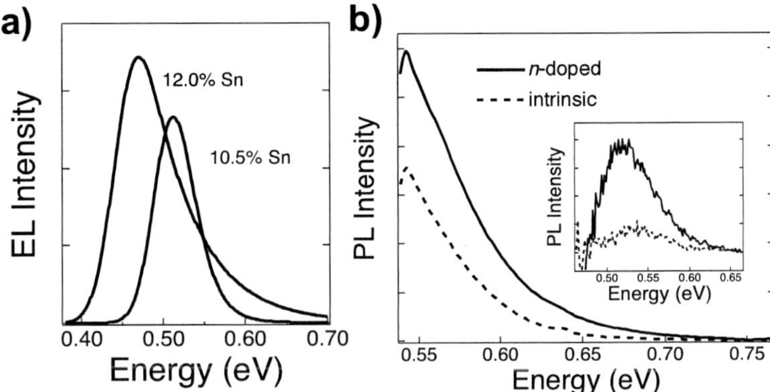

Figure 5 – a) EMG fits to EL spectra obtained from double heterostructure $Ge_{1-y}Sn_y$ *pin* devices with $y = 0.105$ and $y = 0.12$ active layers. b) PL spectra obtained from a liquid nitrogen cooled InGaAs detector for *n*-type doped (solid line) and intrinsic (dashed line) $Ge_{1-y}Sn_y$ samples with $y = 0.09$. Both samples show strong luminescence as expected for direct gap materials. However the doped sample, which has an active carrier concentration of 1×10^{19} cm^{-3}, shows clear enhancement of PL compared to the intrinsic sample. The inset shows the same spectra obtained from a thermoelectrically cooled PbS detector.

In addition to EL, PL from the *n*-type doped layers was also investigated. The results obtained are in keeping with the expectation that the higher electron population in the Γ valley of these samples will enhance their emission compared to intrinsic counterparts. This can be clearly seen in Figure 5 b), which compares the PL obtained from *n*-type and intrinsic $Ge_{1-y}Sn_y$ samples with $y = 0.09$.

Conclusion

In this work we have developed low temperature doping procedures for $Ge_{1-y}Sn_y$ alloys beyond the indirect to direct gap crossover point. These procedures were successfully applied to fabricate *p-i-n* devices which exhibit direct gap EL. Future applications that can be affected by this technology include GeSn emitters and detectors which operate at mid-IR wavelengths, and GeSn laser diodes. Furthermore, enhanced PL was obtained from *n*-type doped $Ge_{1-y}Sn_y$ direct gap alloys.

Acknowledgements

We gratefully acknowledge the use of facilities within the Center for Solid State Science at Arizona State University, including the John M. Cowley Center for High

Resolution Electron Microscopy. This work was funded by AFOSR FA9550-12-1-0208 and AFOSR FA9550-13-1-0022.

References

1. R. Loo, B. Vincent, F. Gencarelli, C. Merckling, A. Kumar, G. Eneman, L. Witters, W. Vandervorst, M. Caymax, M. Heyns, and A. Thean, *ECS J. Solid State Sci. Technol.*, **2**, N35–N40 (2013).
2. B. Vincent, Y. Shimura, S. Takeuchi, T. Nishimura, G. Eneman, a. Firrincieli, J. Demeulemeester, a. Vantomme, T. Clarysse, O. Nakatsuka, S. Zaima, J. Dekoster, M. Caymax, and R. Loo, *Microelectron. Eng.*, **88**, 342–346 (2011).
3. S. Gupta, R. Chen, B. Vincent, D. Lin, and B. Magyari-köpe, *ECS Trans.*, **50**, 937–941 (2012).
4. L. Liu, R. Liang, J. Wang, and J. Xu, *J. Appl. Phys.*, **117**, 184501 (2015).
5. L. Jiang, J. D. Gallagher, C. L. Senaratne, T. Aoki, J. Mathews, J. Kouvetakis, and J. Menéndez, *Semicond. Sci. Technol.*, **115028** (2014).
6. C. L. Senaratne, J. D. Gallagher, T. Aoki, J. Kouvetakis, and J. Menéndez, *Chem. Mater.*, **26**, 6033–6041 (2014).
7. R. W. Olesinski and G. J. Abbaschian, *Bull. Alloy Phase Diagrams*, **5**, 265–271 (1984).
8. M. Oehme, J. Werner, M. Gollhofer, M. Schmid, M. Kaschel, E. Kasper, and J. Schulze, **23**, 19–21 (2011).
9. H. H. Tseng, H. Li, V. Mashanov, Y. J. Yang, H. H. Cheng, G. E. Chang, R. A. Soref, and G. Sun, *Appl. Phys. Lett.*, **103**, 231907 (2013).
10. W. Du, Y. Zhou, S. A. Ghetmiri, A. Mosleh, B. R. Conley, A. Nazzal, R. A. Soref, G. Sun, J. Tolle, J. Margetis, H. a. Naseem, and S.-Q. Yu, *Appl. Phys. Lett.*, **104**, 241110 (2014).
11. J. P. Gupta, N. Bhargava, S. Kim, T. Adam, and J. Kolodzey, *Appl. Phys. Lett.*, **102**, 2011–2015 (2013).
12. H. H. Tseng, K. Y. Wu, H. Li, V. Mashanov, H. H. Cheng, G. Sun, and R. A. Soref, *Appl. Phys. Lett.*, **102**, 182106 (2013).
13. A. V. G. Chizmeshya, C. Ritter, J. Tolle, C. Cook, J. Menéndez, and J. Kouvetakis, *Chem. Mater.*, **18**, 6266–6277 (2006).
14. J. Xie, J. Tolle, V. R. D'Costa, C. Weng, a. V. G. Chizmeshya, J. Menendez, and J. Kouvetakis, *Solid. State. Electron.*, **53**, 816–823 (2009).
15. J. D. Gallagher, C. L. Senaratne, P. Sims, T. Aoki, J. Menéndez, and J. Kouvetakis, *Appl. Phys. Lett.*, **106**, 091103 (2015).
16. J. D. Gallagher, C. L. Senaratne, C. Xu, P. Sims, T. Aoki, D. J. Smith, J. Menéndez, and J. Kouvetakis, *J. Appl. Phys.*, **117**, 245704 (2015).
17. C. Xu, R. T. Beeler, L. Jiang, G. Grzybowski, A. V. G. Chizmeshya, J. Menéndez, and J. Kouvetakis, *Semicond. Sci. Technol.*, **28**, 105001 (2013).

ECS Transactions, 69 (14) 165-166 (2015)
10.1149/06914.0165ecst ©The Electrochemical Society

Author Index